GLENCOE MATH

Interactive Guide

CONTRIBUTING AUTHORS

Philip Gonsalves
Director of Curriculum and Instruction for Mathematics
West Contra Costa Unified School District
Richmond, California

Dinah Zike
Educational Consultant
Dinah-Might Activities, Inc.
San Antonio, Texas

Mc
Graw
Hill
Education

Bothell, WA • Chicago, IL • Columbus, OH • New York, NY

mheducation.com/prek-12

Send all inquiries to:
McGraw-Hill Education
STEM Learning Solutions Center
8787 Orion Place
Columbus, OH 43240

ISBN: 978-0-07-898939-1
MHID: 0-07-898939-6

Printed in the United States of America.

1 2 3 4 5 6 7 8 9 LWI 22 21 20 19 18 17

Visual Kinesthetic Vocabulary® is a registered trademark of
Dinah-Might Adventures, LP.

Contents

Chapter 9 Area

Chapter 10 Volume and Surface Area

Chapter 11 Statistical Measures

Chapter 12 Statistical Displays

Lesson 1 Notetaking

Factors and Multiples

Use Cornell notes to better understand the lesson's concepts. Complete each sentence by filling in the blanks with the correct word or phrase.

Questions	Notes
1. How do I find the greatest common factor?	I can use _____ to find the greatest of the common factors of two or more numbers. This is called the _____ .
2. How do I find the least common multiple?	I can use a _____, _____, or _____ to find the _____ whole number that is a common multiple of each of two or more numbers. This is called the _____ .

Summary

Describe the differences between the least common multiple and the greatest common factor.

_____

_____

_____

_____

_____

_____

_____

Inquiry Lab Guided Writing

Ratios

HOW can you use tables to relate quantities?

Use the exercises below to help answer the Inquiry Question. Write the correct word or phrase on the lines provided.

1. Rewrite the question in your own words.

   _____

   _____

2. What key words do you see in the question?

   _____

3. *Quantity* is another word for "number." What is another word for *quantities*?

   _____

4. What is the product of a number and any whole number?

   _____

5. A _____ is a graphic organizer that displays data in rows and columns.

6. Write the missing multiples in the table below.

2	4	6		10	
3	6		12		18

7. What common multiples of 2 and 3 does the table show?

   _____

   HOW can you use tables to relate quantities?

   _____

   _____

Lesson 2 Vocabulary
Ratios

Use the concept web to show the ratio of 2 to 3 in different ways. Use a diagram in one of the pieces of the web.

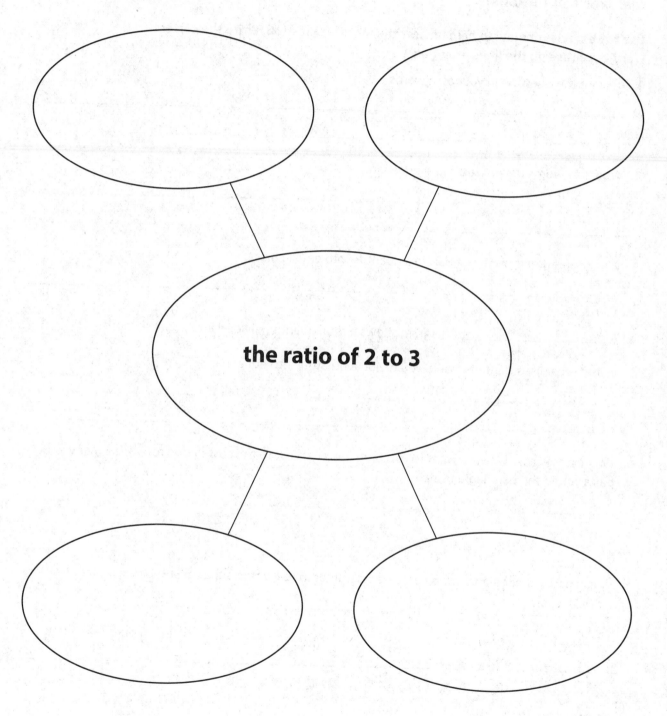

Inquiry Lab Guided Writing

Unit Rates

HOW can you use bar diagrams to compare quantities in real-world situations?

Use the exercises below to help answer the Inquiry Question. Write the correct word or phrase on the lines provided.

1. Rewrite the question in your own words.

_____

_____

2. What key words do you see in the question?

_____

_____

3. What operation is used to separate a quantity into equal parts?

_____

4. A _____ diagram shows a total quantity divided into equal sections.

5. In what real-life situations would you divide a quantity into equal parts?

_____

_____

6. Twelve roses cost $30. You want to compare the total cost to the cost of one rose. How many equal parts would the bar diagram show?

_____

HOW can you use bar diagrams to compare quantities in real-world situations?

_____

_____

_____

Lesson 3 Vocabulary
Rates

Use the vocabulary squares to write a definition, a sentence, and an example for each vocabulary word.

ratio	Definition
Example	**Sentence**

rate	Definition
Example	**Sentence**

unit rate	Definition
Example	**Sentence**

Lesson 4 Notetaking
Ratio Tables

Use Cornell notes to better understand the lesson's concepts. Complete each sentence by filling in the blanks with the correct word or phrase.

Questions	Notes
1. What are equivalent ratios?	Equivalent ratios are ratios that express _____ between quantities.
2. How can I use scaling to find equivalent ratios?	By multiplying or dividing two related quantities by _____, I can scale forward or scale back.

Summary

How can I determine if two ratios are equivalent?

_____

_____

_____

_____

_____

_____

_____

Lesson 5 Vocabulary
Graph Ratio Tables

Use the concept web to identify the parts of the coordinate plane. Use the words from the word bank.

Word Bank

origin	x-axis	y-axis
ordered pair	x-coordinate	y-coordinate

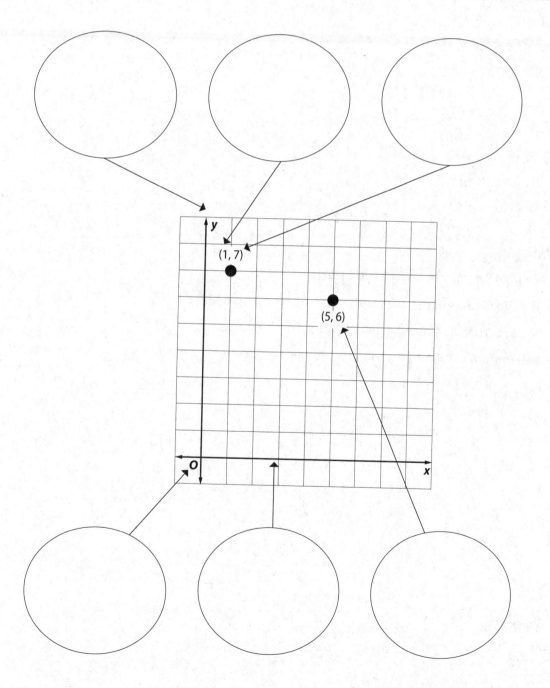

Problem-Solving Investigation
The Four-Step Plan

Case 3 Walking

Megan uses a pedometer to find how many steps she takes each school day.

She took **32,410 steps** over the course **of 5 days**.

If she took the **same number of steps each day, how many did she take on Monday**?

- Understand:

- Plan:

- Solve:

- Check:

Case 4 Savings

James is earning money to buy a **$100 bicycle**.

For **each dollar James earns, his mother** has agreed to **give him $1**.

So far, he has earned **$14** mowing lawns and **$7** washing cars.

How much **more** must James **earn** in order to buy the bicycle?

- Understand:

- Plan:

- Solve:

- Check:

Lesson 6 Review Vocabulary
Equivalent Ratios

Use the definition map to list qualities about the vocabulary word or phrase.

Vocabulary

equivalent ratios

**Characteristics
(What is it like?)**

Description

Examples

Inquiry Lab Guided Writing

Ratio and Rate Problems

HOW can you use unit rates and multiplication to solve for missing measures in equivalent ratio problems?

Use the exercises below to help answer the Inquiry Question. Write the correct word or phrase on the lines provided.

1. Rewrite the question in your own words.

   _____

   _____

2. What key words do you see in the question?

   _____

   _____

3. Use the bar diagram and exercises to find the number of carrots that can be chopped in 10 minutes at a rate of 6 carrots per 2 minutes.

   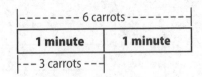

   a. What is the initial rate that is given? _____ carrots in _____ minutes

   b. What kind of model is used to find the unit rate? _____

   c. What is the unit rate? _____ carrots in _____ minute

   d. The missing measure is the number of carrots chopped in _____ minutes.

   e. What operation will you perform on the unit rate to find the missing measure?

      _____

   f. Complete the equation to find the missing measure.

      _____ $\dfrac{\text{carrots}}{\text{minute}} \times$ _____ minutes = _____ carrots

   HOW can you use unit rates and multiplication to solve for missing measures in equivalent ratio problems?

   _____

   _____

Lesson 7 Review Vocabulary
Ratio and Rate Problems

Use the word cards to define each vocabulary word or phrase and give an example.

```
                        Word Cards

┌─────────────────────────┐    ┌─────────────────────────┐
│          ratio          │    │          razón          │
└─────────────────────────┘    └─────────────────────────┘

Definition                     Definición

_____        _____
_____        _____
_____        _____

Example Sentence

_____

_____
```

- -

```
                        Word Cards

┌─────────────────────────┐    ┌─────────────────────────┐
│          rate           │    │          tasa           │
└─────────────────────────┘    └─────────────────────────┘

Definition                     Definición

_____        _____
_____        _____
_____        _____

Example Sentence

_____

_____
```

NAME _____ DATE _____ PERIOD _____

Lesson 1 Vocabulary
Decimals and Fractions

Use the concept web to give examples of rational numbers.

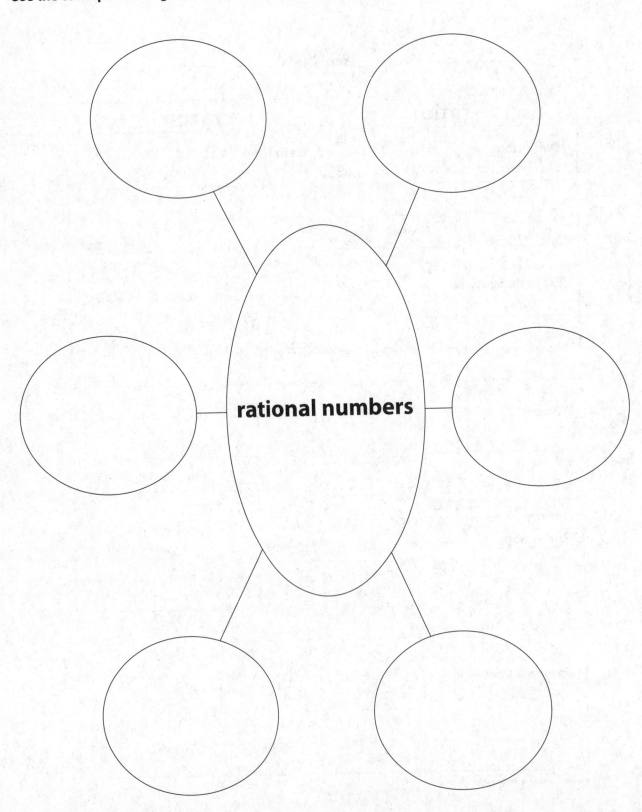

Inquiry Lab Guided Writing

Model Percents

HOW can you model a percent?

Use the exercises below to help answer the Inquiry Question. Write the correct word or phrase on the lines provided.

1. Rewrite the question in your own words.

_____

_____

2. What key words do you see in the question?

_____

3. The word _____ means "out of one hundred."

4. How many squares are in a 10 × 10 grid? _____

5. Shade 65 squares in the 10 × 10 grid below.

6. A 10 × 10 grid with 65 squares shaded is a model for _____ percent.

7. What percent is modeled by the bar diagram below? _____

0%   100%
| 10% | 10% | 10% | 10% | 10% | 10% | 10% | 10% | 10% | 10% |

8. What other percents would be easy to model on a bar diagram?

_____

HOW can you model a percent?

_____

_____

_____

Lesson 2 Vocabulary
Percents and Fractions

Use the definition map to list qualities about the vocabulary word or phrase.

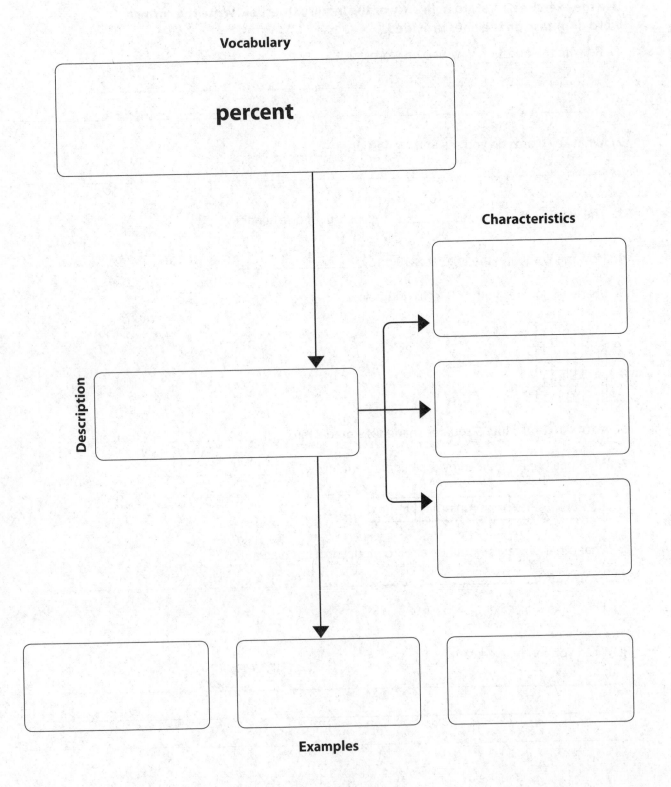

Vocabulary

percent

Description

Characteristics

Examples

Lesson 3 Notetaking
Percents and Decimals

Use Cornell notes to better understand the lesson's concepts. Complete each sentence by filling in the blanks with the correct word or phrase.

Questions	Notes
1. How do I write percents as decimals?	Divide by _____ and remove the _____. This is the same as moving the decimal point _____ places to the _____.
2. How do I write decimals as percents?	Multiply by _____ and add a _____. This is the same as moving the decimal point _____ places to the _____.

Summary

What is the relationship between percents and decimals?

_____

_____

_____

_____

_____

_____

_____

Lesson 4 Review Vocabulary

Percents Greater than 100% and Percents Less than 1%

Use the word cards to define each vocabulary word or phrase and give an example.

Word Cards

percent	por ciento
Definition	**Definición**
_____	_____
_____	_____
_____	_____
Example Sentence	
_____	
_____	

- -

Word Cards

fraction	fracción
Definition	**Definición**
_____	_____
_____	_____
_____	_____
Example Sentence	
_____	
_____	

Problem-Solving Investigation
Solve a Simpler Problem

Case 3 Time

Three 24-hour **clocks** show the time to be 12 noon.

One of the clocks **is always correct, one loses a minute every 24 hours**, and **one gains a minute every 24 hours**.

How many hours will pass before all three clocks show the correct time again?

- Understand:

- Plan:

- Solve:

- Check:

Case 4 Number Sense

The number **272 is a palindrome** because it **reads the same forward or backward.**

How many numbers from 10 to 1,000 are palindromes?

- Understand:

- Plan:

- Solve:

- Check:

Lesson 5 Vocabulary

Compare and Order Fractions, Decimals, and Percents

Use the definition map to list qualities about the vocabulary word or phrase.

Vocabulary

least common denominator (LCD)

Description

Characteristics (When is it used?)

Examples

Lesson 6 Notetaking

Estimate with Percents

Use Cornell notes to better understand the lesson's concepts. Complete each sentence by filling in the blanks with the correct word or phrase.

Questions	Notes
1. How can I estimate the percent of a number?	Round the percent and the number using _____. Write the rounded percent as a fraction. _____ the fraction and the rounded number to find the _____.
2. How can I estimate using the rate per 100?	Write the percent as a rate per 100. Round the number to the nearest _____. Then _____ the rate by the number of hundreds.

Summary

What is the relationship between percents and decimals?

_____

_____

_____

_____

_____

_____

_____

_____

Inquiry Lab Guided Writing

Percent of a Number

HOW can you model the percent of a number?

Use the exercises below to help answer the Inquiry Question. Write the correct word or phrase on the lines provided.

1. Rewrite the question in your own words.

   _____

   _____

2. What key words do you see in the question?

   _____

   _____

3. What models have you used to represent percents?

   _____

4. Which model is easy to divide into equal parts?

   _____

5. Suppose you want to find 75% of 24 dogs. Label what each bar diagram below shows.

25%	25%	25%	25%	100%

_____

6	6	6	6

_____

6. How many bar diagrams are needed to model the percent of a number? _____

7. Are the bar diagrams the same length or different lengths? _____

8. Are both bar diagrams divided into equal parts? _____

HOW can you model the percent of a number?

_____

_____

_____

Lesson 7 Review Vocabulary

Percent of a Number

Use the concept web to write examples of twenty-five percent. Use a diagram in at least one part of the web.

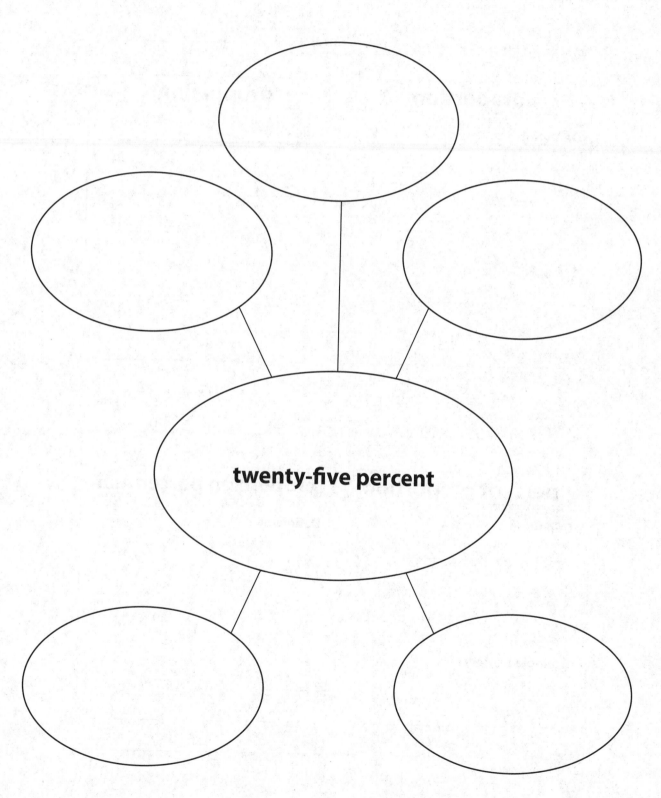

twenty-five percent

Lesson 8 Vocabulary

Solve Percent Problems

Use the word cards to define each vocabulary word or phrase and give an example.

Word Cards

proportion	proporción
Definition	**Definición**
_____	_____
_____	_____
_____	_____
Example Sentence	
_____	
_____	

- -

Word Cards

percent proportion	proporción porcentual
Definition	**Definición**
_____	_____
_____	_____
_____	_____
Example Sentence	
_____	
_____	

Lesson 1 Notetaking

Add and Subtract Decimals

Use Cornell notes to better understand the lesson's concepts. Complete each sentence by filling in the blanks with the correct word or phrase.

Questions	Notes
1. How do I add decimals?	To add decimals, I line up the _____ . Then, add digits in the same _____ .
2. How do I subtract decimals?	To subtract decimals, I line up the _____ . Then, subtract digits in the same _____ . I may need to _____ , or place _____ at the end of a decimal in order to subtract.

Summary

How is estimation helpful when adding and subtracting decimals?

_____

_____

_____

_____

_____

_____

_____

_____

Lesson 2 Review Vocabulary

Estimate Products

Use the word cards to define each vocabulary word or phrase and give an example.

Word Cards

estimate

Definition

_____

_____

_____

Example Sentence

_____

_____

estimar

Definición

_____

_____

_____

Word Cards

product

Definition

_____

_____

_____

Example Sentence

_____

_____

producto

Definición

_____

_____

_____

Lesson 3 Review Vocabulary
Multiply Decimals by Whole Numbers

Complete the four-square chart to review the word or phrase. Then answer the question below.

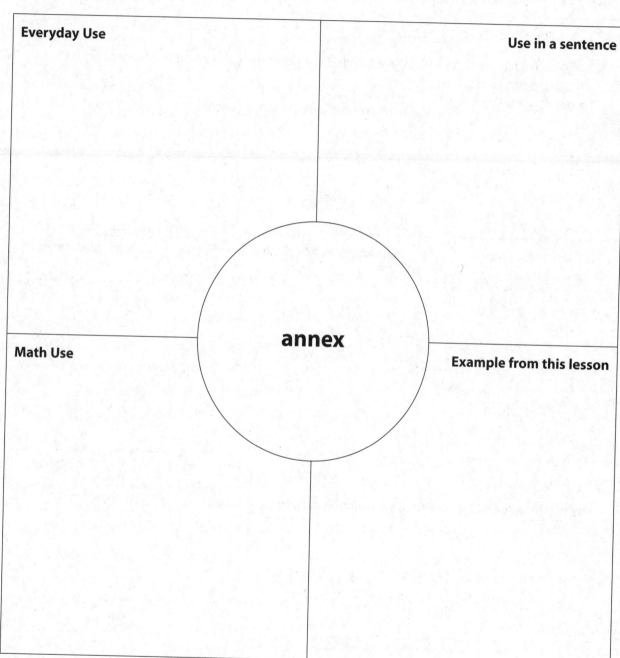

Everyday Use	Use in a sentence
Math Use	**Example from this lesson**

annex

What does it mean to annex a zero?

To annex a zero means to place a _____ at the_____

or the _____ of a decimal.

Lesson 4 Notetaking

Multiply Decimals by Decimals

Use Cornell notes to better understand the lesson's concepts. Complete each sentence by filling in the blanks with the correct word or phrase.

Questions	Notes
1. How do I place the decimal point in the product of two decimals?	Place the decimal point the same number of places from the _____ as the _____ of the number of decimal places in each factor.
2. When would I annex zeros in a product?	I annex a zero when there are not enough decimal places in the _____. I annex zeros to the _____.

Summary
Why is estimating not as helpful when multiplying very small numbers such as 0.007 and 0.053?

_____

_____

_____

_____

_____

_____ |

Inquiry Lab Guided Writing

Multiply by Powers of Ten

HOW can number patterns be used to multiply by powers of 10?

Use the exercises below to help answer the Inquiry question. Write the correct word or phrase on the lines provided.

1. Rewrite the question in your own words.

_____

_____

2. What key words do you see in the question?

_____

3. A predictable relationship among numbers is called a _____.

4. Name three different powers of 10. _____

5. Look at the multiplication sentence: $14 \times 100{,}000 = 1{,}400{,}000$.
What is the same about the power of 10 and the product?

_____

_____

6. Look at the multiplication sentence: $14 \times 0.001 = 0.014$.
What is the same about the power of 10 and the product?

_____

_____

7. To multiply by a power of 10 greater than 1, count the _____.

8. To multiply by a power of 10 less than 1, count the _____.

HOW can number patterns be used to multiply by powers of 10?

_____

_____

Problem-Solving Investigation
Look for a Pattern

Case 3 Gaming

The table shows the cost of a subscription to the Action Gamers Channel.

What is the **cost of a 1 year subscription**?

- Understand:

- Plan:

- Solve:

Action Gamers Channel Prices	
Number of Months	Total Cost($)
2	15.90
3	23.85
4	31.80

- Check:

Case 4 Number Theory

The diagram to the right is known as Pascal's Triangle.

If the pattern continues, **what will the numbers in the next row be** from left to right?

- Understand:

- Plan:

- Solve:

- Check:

Lesson 5 Review Vocabulary
Divide Multi-Digit Numbers

Use the concept web to identify the parts of a division problem. Use the terms from the word bank.

Word Bank			
dividend	divisor	quotient	remainder

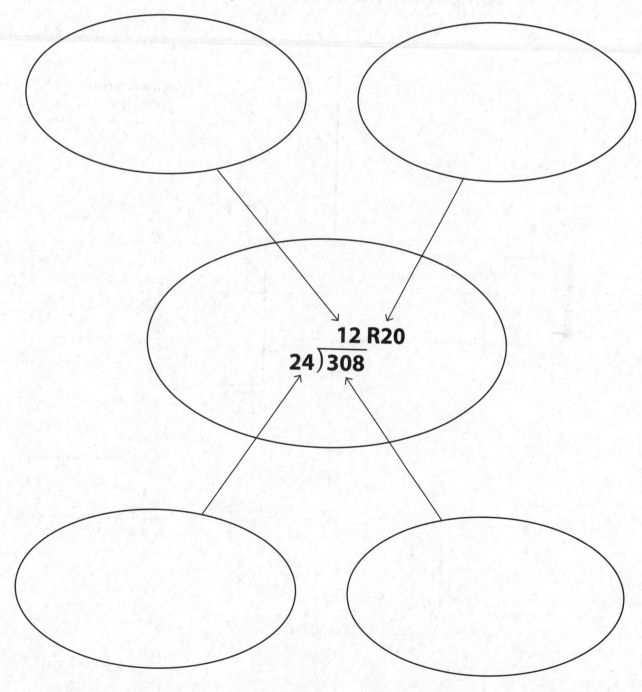

$$12\ \text{R}20$$
$$24\overline{)308}$$

Lesson 6 Vocabulary

Estimate Quotients

Use the definition map to list qualities about the vocabulary word or phrase.

Vocabulary

compatible numbers

Description

**Characteristics
(What is it like?)**

Examples

Lesson 7 Notetaking
Divide Decimals by Whole Numbers

Use Cornell notes to better understand the lesson's concepts. Complete each sentence by filling in the blanks with the correct word or phrase.

Questions	Notes
1. Where do I place the decimal point in the quotient when a decimal is divided by a whole number?	The decimal point in the quotient is placed directly _____ the decimal point in the _____ .
2. Why would I annex a zero during division?	Annexing a zero to the _____ allows me to continue dividing without changing the value of the _____ .

Summary

How can estimating quotients help me to place the decimal correctly?

_____

_____

_____

_____

_____

_____

_____

Lesson 8 Notetaking

Divide Decimals by Decimals

Use the K-W-L chart to better understand the lesson's concept on division. Fill in the chart with the correct word or phrase.

K - What I Already Know

From experience	From previewing the lesson

W - What I Want to Learn

Look for the Building on the Essential Question. Rewrite it here.

L - What I Learned

Show examples of division problems with decimals.

Lesson 1 Review Vocabulary

Estimate Products of Fractions

Use the definition map to list qualities about the vocabulary word or phrase.

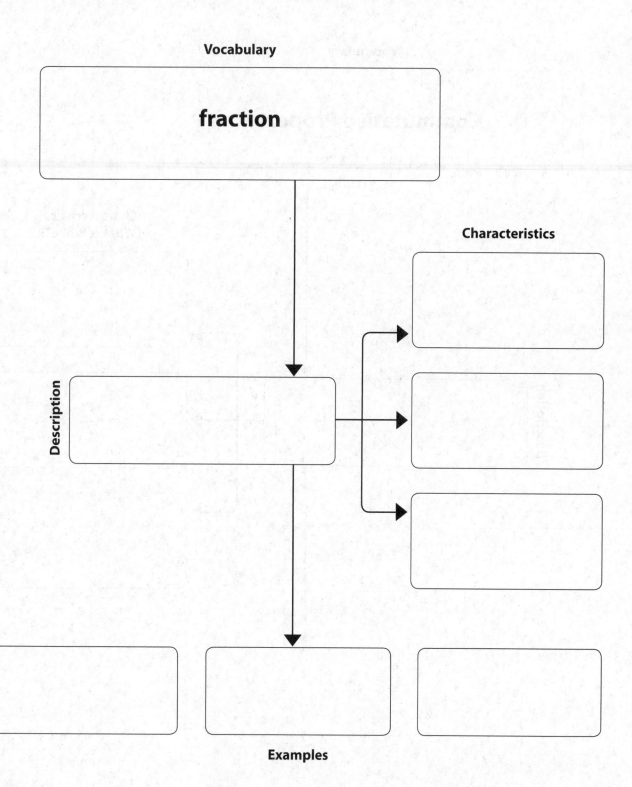

Vocabulary

fraction

Characteristics

Description

Examples

Lesson 2 Vocabulary

Multiply Fractions and Whole Numbers

Use the definition map to list qualities about the vocabulary word or phrase.

Vocabulary

Commutative Property

**Characteristics
(What is it like?)**

Description

Examples

Lesson 3 Notetaking
Multiply Fractions

Use Cornell notes to better understand the lesson's concepts. Complete each sentence by filling in the blanks with the correct word or phrase.

Questions	Notes
1. How do I multiply two fractions?	Multiply the _____ and then multiply the _____ .
2. How do I simplify before multiplying fractions?	I can simplify before I multiply, if _____ and _____ have a common factor. _____ both the numerator and denominator by the common factor.

Summary
If two positive fractions are less than 1, why is their product also less than 1?
_____
_____
_____
_____
_____
_____
_____
_____

Lesson 4 Review Vocabulary
Multiply Mixed Numbers

Use the word cards to define each vocabulary word or phrase and give an example.

Word Cards

mixed number	número mixto
Definition	**Definición**
_____	_____
_____	_____
_____	_____
Give 3 examples.	
_____	_____
_____	_____

- -

Word Cards

improper fraction	fracción impropia
Definition	**Definición**
_____	_____
_____	_____
_____	_____
Give 3 examples.	
_____	_____
_____	_____

Lesson 5 Vocabulary
Convert Measurement Units

Use the word Cards to define each vocabulary word or phrase and give an example.

Word Cards

unit ratio

Definition

_____

_____

_____

Example Sentence

_____

_____

razón unitaria

Definición

_____

_____

_____

- -

Word Cards

dimensional analysis

Definition

_____

_____

_____

Example Sentence

_____

_____

análisis dimensional

Definición

_____

_____

_____

Problem-Solving Investigation
Draw a Diagram

Case 3 Internet

Francesca spent **25 minutes on the Internet** yesterday.

If this is $\frac{5}{6}$ **of the time she spent on the computer**, how long did she spend on the computer, but not on the Internet?

- Understand:

- Plan:

- Solve:

- Check:

Case 4 Basketball

Meiko practiced **shooting** a basketball for $\frac{7}{10}$ **of her total practice time**. During the other time, she practiced dribbling.

If she practiced **dribbling for 18 minutes**, how many minutes did she practice shooting?

- Understand:

- Plan:

- Solve:

- Check:

Inquiry Lab Guided Writing

Divide Whole Numbers by Fractions

HOW can a bar diagram help you understand what it means to divide fractions?

Use the exercises below to help answer the Inquiry Question. Write the correct word or phrase on the lines provided.

1. Rewrite the question in your own words.

_____

_____

2. What key words do you see in the question?

_____

Use the equation $6 \div \frac{2}{3} = 9$ and the model shown to answer Exercises 3-7.

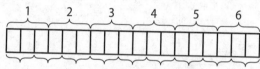

3. The bar diagram shows the dividend (6) divided into equal groups of _____ .

4. The bar diagram shows that there are _____ groups of $\frac{2}{3}$.

5. The fraction $\frac{2}{3}$ is the divisor. Is $\frac{2}{3}$ greater than or less than 1? _____

6. Is the quotient greater than or less than the dividend? _____

7. So, the model shows that when the divisor is _____ than 1, the quotient is _____ than the dividend.

HOW can a bar diagram help you understand what it means to divide fractions?

_____

_____

_____

Lesson 6 Vocabulary
Divide Whole Numbers by Fractions

Use the definition map to list qualities about the vocabulary word or phrase.

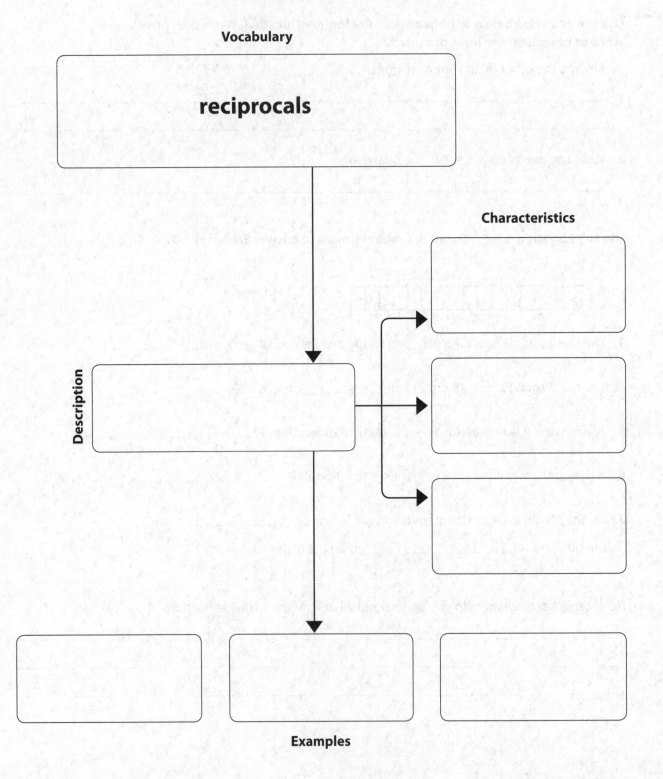

Vocabulary

reciprocals

Characteristics

Description

Examples

Inquiry Lab Guided Writing

Divide Fractions

HOW can using models help you divide one fraction by another fraction?

Use the exercises below to help answer the Inquiry Question. Write the correct word or phrase on the lines provided.

1. Rewrite the question in your own words.

_____

_____

2. What key words do you see in the question?

_____

Use the equation $\frac{4}{5} \div \frac{2}{5} = 2$ and the model shown to answer Exercises 3-7.

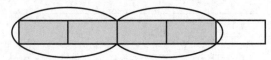

3. The shaded part of the bar diagram represents the dividend. That is the fraction _____ .

4. The circled parts represent the divisor. That is the fraction _____ .

5. Are the fractions $\frac{4}{5}$ and $\frac{2}{5}$ greater than or less than 1? _____

6. Is the quotient greater than 1? _____

7. The model shows that when the dividend is greater than the divisor, the quotient is

_____ than 1.

HOW can using models help you divide one fraction by another fraction?

_____

_____

_____

Lesson 7 Notetaking
Divide Fractions

Use Cornell notes to better understand the lesson's concepts. Complete each sentence by filling in the blanks with the correct word or phrase.

Questions	Notes
1. How do I divide a fraction by a fraction?	Rewrite the division expression as a _____ expression, multiplying by the reciprocal of the _____ .
2. How do I divide a fraction by a whole number?	Write the _____ as a _____ . Then _____ as with fractions.

Summary
How is the process used to divide fractions similar to the process used to multiply fractions?
_____
_____
_____
_____
_____
_____

Lesson 8 Notetaking

Divide Mixed Numbers

Use Cornell notes to better understand the lesson's concepts. Complete each sentence by filling in the blanks with the correct word or phrase.

Questions	Notes
1. How do I divide a mixed number by a fraction?	First write the _____ as _____. Then _____ as with fractions.
2. How do I simplify before dividing mixed numbers?	Write each _____ as an _____ . Then rewrite the expression so you multiply by the reciprocal. Find the greatest common factor of the _____ and the _____ . _____ both the numerator and denominator by the greatest common factor.

Summary
How do I divide mixed numbers?
_____
_____
_____
_____
_____
_____
_____
_____

Inquiry Lab Guided Writing

Integers

HOW can positive and negative values be represented?

Use the exercises below to help answer the Inquiry Question. Write the correct word or phrase on the lines provided.

1. Rewrite the question in your own words.

_____

_____

2. What key words do you see in the question?

_____

3. Is a positive number greater than or less than zero? _____

4. Does the (+) symbol represent a positive or negative number? _____

5. Is a negative number greater than or less than zero? _____

6. Does the (−) symbol represent a positive or negative number? _____

7. Write the missing numbers on the number line below.

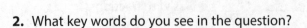

−4 _____ −2 _____ 0   1 _____ _____ 4

8. What negative numbers did you write? _____

9. What positive numbers did you write? _____

HOW can positive and negative values be represented?

_____

_____

_____

Lesson 1 Vocabulary
Integers and Graphing

Use the vocabulary squares to write a definition, a sentence, and an example for each vocabulary word.

	Definition
integer	
Example	**Sentence**

	Definition
positive integer	
Example	**Sentence**

	Definition
negative integer	
Example	**Sentence**

Inquiry Lab Guided Writing

Absolute Value

HOW can a number line help you find two integers that are the same distance from zero?

Use the exercises below to help answer the Inquiry Question. Write the correct word or phrase on the lines provided.

1. Rewrite the question in your own words.

_____

_____

2. What key words do you see in the question?

_____

Use the number line below to answer Exercises 3–8.

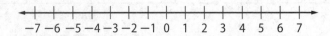

$$-7\ -6\ -5\ -4\ -3\ -2\ -1\ \ 0\ \ 1\ \ 2\ \ 3\ \ 4\ \ 5\ \ 6\ \ 7$$

3. Are positive numbers to the right or left of zero? _____

4. Are negative numbers to the right or left of zero? _____

5. A(n) _____ is a positive or negative whole number.

6. How many spaces from zero is the integer 2? _____

7. What other integer is 2 spaces from zero? _____

8. What two integers are 7 spaces from zero? _____

HOW can a number line help you find two integers that are the same distance from zero?

_____

_____

Lesson 2 Vocabulary
Absolute Value

Use the Word Cards to define each vocabulary word or phrase and give an example.

Word Cards

absolute value	**valor absoluto**
Definition	**Definición**
_____	_____
_____	_____
_____	_____
Example Sentence	
_____	
_____	

- -

Word Cards

opposites	**opuestos**
Definition	**Definición**
_____	_____
_____	_____
_____	_____
Example Sentence	
_____	
_____	

Lesson 3 Notetaking
Compare and Order Integers

Use Cornell notes to better understand the lesson's concepts. Complete each sentence by filling in the blanks with the correct word or phrase.

Questions	Notes
1. How do I compare integers?	I can compare signs. _____ numbers are greater than _____ numbers. I can compare position on the number line. Greater numbers are graphed farther to the _____ .
2. How do I order integers?	I can use a _____ to order a set of integers. I can compare _____ and _____ to order a set of integers.

Summary

How can symbols and absolute value help you to order sets of integers?

_____

_____

_____

_____

_____

_____

Problem-Solving Investigation
Work Backward

Case 3 Sea Level

Mr. Ignacio went diving along the coral reef in Oahu.

He **descended 12 meters below sea level**.

The **difference between this** point on the coral reef **and the highest point** on the island, Mount Ka'ala, is **1,232 meters**.

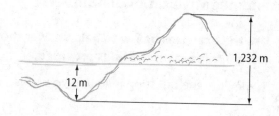

Two-fifths of the way up the mountain is a ranger station. How far above sea level is the ranger station?

- Understand:

- Plan:

- Solve:

- Check:

Case 4 Cameras

Adamo **saved 13 pictures** on his digital camera for a **total of 12,021.1 KB**.

He **deleted 32 pictures** for a **total of 29,590.4 KB**.

If there are **now 108 pictures**, how many kilobytes of storage did he use at the beginning?

- Understand:

- Plan:

- Solve:

- Check:

Inquiry Lab Guided Writing

Number Lines

HOW can you use a number line to model and compare positive and negative rational numbers?

Use the exercises below to help answer the Inquiry question. Write the correct word or phrase on the lines provided.

1. Rewrite the question in your own words.

_____

_____

2. Decimals and fractions are _____.

3. On a number line, numbers to the left are _____ than numbers to the right.

Use the number line below for Exercises 4-9.

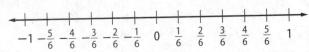

4. Which number is smaller, $-\frac{1}{6}$ or $-\frac{5}{6}$? _____

5. Which fraction is farther from 0, $-\frac{1}{6}$ or $-\frac{5}{6}$? _____

6. Which number is greater, $\frac{2}{6}$ or $\frac{4}{6}$? _____

7. Which fraction is farther from 0, $\frac{2}{6}$ or $\frac{4}{6}$? _____

8. Compare two negative rational numbers on the number line. Use < or >. _____

9. Compare two positive rational numbers on the number line. Use < or >. _____

HOW can you use a number line to model and compare positive and negative rational numbers?

_____

_____

_____

Lesson 4 Vocabulary
Terminating and Repeating Decimals

Use the vocabulary squares to write a definition, a sentence, and an example for each vocabulary word.

rational number	Definition
Example	**Sentence**

terminating decimal	Definition
Example	**Sentence**

repeating decimal	Definition
Example	**Sentence**

Lesson 5 Notetaking

Compare and Order Rational Numbers

Use Cornell notes to better understand the lesson's concepts. Complete each sentence by filling in the blanks with the correct word or phrase.

Questions	Notes
1. How do I compare and order two fractions?	If the fractions do not have the same _____, I must _____ the fractions using the least common denominator. Then I can use a _____ to compare and order the two fractions.
2. How do I compare and order rational numbers?	First write the rational numbers in the same _____. Then I can use a _____ to compare and order the numbers.

Summary
How can a number line help in ordering rational numbers?
_____
_____
_____
_____
_____
_____
_____

Lesson 6 Vocabulary

The Coordinate Plane

Use the concept web to identify the quadrants of the coordinate plane. Write an ordered pair to name a point in each quadrant.

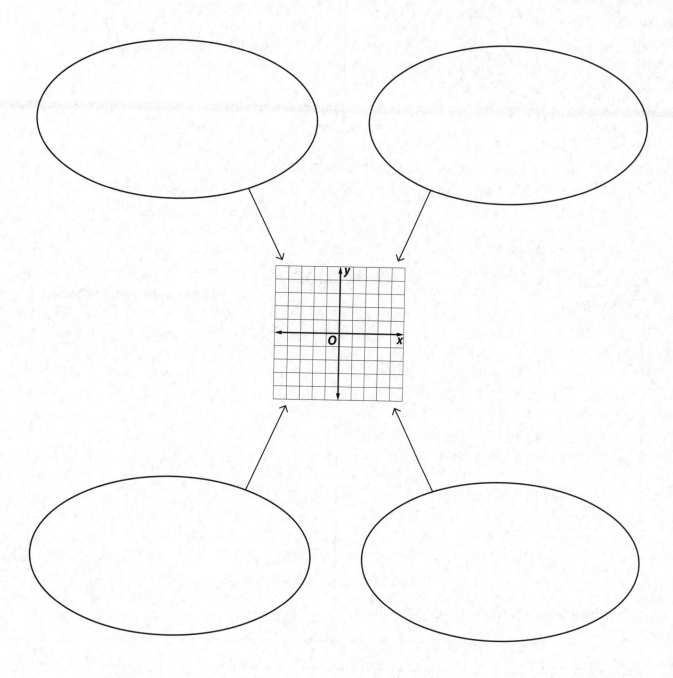

Lesson 7 Review Vocabulary
Graph on the Coordinate Plane

Complete the four-square chart to review the word or phrase. Then answer the question below.

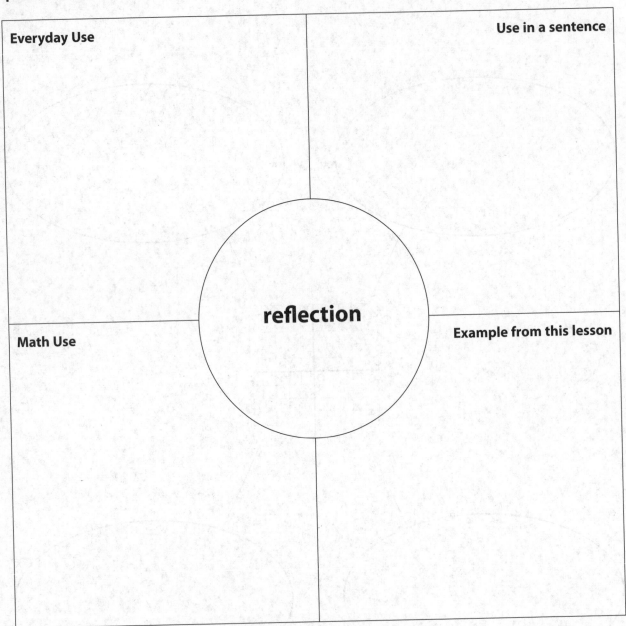

Everyday Use	Use in a sentence
Math Use	**Example from this lesson**

reflection

What does it mean to reflect a point across the *x*-axis?

_____

_____

_____

NAME _____ DATE _____ PERIOD _____

Inquiry Lab Guided Writing

Find Distance on the Coordinate Plane

WHAT is the relationship between coordinates and distance?

Use the exercises below to help answer the Inquiry Question. Write the correct word or phrase on the lines provided.

1. Rewrite the question in your own words.

_____

_____

2. What key words do you see in the question?

_____

Use the coordinate plane below to answer Exercises 3-6.

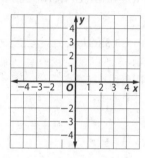

3. Are the *x*-coordinates on a *horizontal* or *vertical* line? _____

4. Are the *y*-coordinates on a *horizontal* or *vertical* line? _____

5. What is the distance between the *x*-coordinates –3 and 2? _____

6. What is the distance between the *y*-coordinates 1 and –2? _____

WHAT is the relationship between coordinates and distance?

_____

_____

_____

Inquiry Lab Guided Writing

Structure of Expressions

HOW can you identify the parts of an expression using mathematical terms?

Use the exercises below to help answer the Inquiry Question.
Write the correct word or phrase on the lines provided.

1. Rewrite the question in your own words.

_____

_____

2. What key words do you see in the question?

_____

_____

3. An _____ is a combination of numbers and operations.

4. Each number or letter in an expression is called a _____ .

5. What operation does each symbol represent?

  a. + _____   **c.** × _____ .

  b. − _____   **d.** ÷ _____

6. The answer to an addition problem is called the _____ .

7. The answer to a subtraction problem is called the _____ .

8. The answer to a multiplication problem is called the _____ .

9. The answer to a division problem is called the _____ .

HOW can you identify the parts of an expression using mathematical terms?

_____

_____

_____

Lesson 1 Vocabulary
Powers and Exponents

Use the three column chart to organize the vocabulary in this lesson. Write the word in Spanish. Then write the definition of each word.

English	Spanish	Definition
base		
exponent		
powers		
perfect square		

Lesson 2 Vocabulary
Numerical Expressions

Use the word cards to define each vocabulary word or phrase and give an example.

Word Cards

numerical expression

Definition

_____

_____

_____

Example Sentence

_____

_____

expressión numérica

Definición

_____

_____

_____

Word Cards

order of operations

Definition

_____

_____

_____

Example Sentence

_____

_____

orden de las operaciones

Definición

_____

_____

_____

Lesson 3 Vocabulary

Algebra: Variables and Expressions

Use the three column chart to organize the vocabulary in this lesson. Write the word in Spanish. Then write the definition of each word.

English	Spanish	Definition
algebra		
variable		
algebraic expression		
evaluate		

Inquiry Lab Guided Writing

Write Expressions

HOW can bar diagrams help you to write expressions in which letters stand for numbers?

Use the exercises below to help answer the Inquiry Question. Write the correct word or phrase on the lines provided.

1. Rewrite the question in your own words.

_____

_____

2. What key words do you see in the question?

_____

Use the expression $h - 5$ and the bar diagram shown to answer Exercises 3-7.

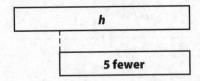

3. Letters stand for numbers that are unknown. What letter is used in the

expression? _____

4. What number is given in the expression? _____

5. Why is the second bar in the bar diagram shorter?

_____

6. Write an expression that can be represented by the bar diagram. _____

7. How does the expression you wrote in Exercise 6 compare to the given expression?

_____

HOW can bar diagrams help you to write expressions in which letters stand for numbers?

_____

_____

Lesson 4 Notetaking

Algebra: Write Expressions

Use Cornell notes to better understand the lesson's concepts. Complete each sentence by filling in the blanks with the correct word or phrase.

Questions	Notes
1. How do I write phrases as algebraic expressions?	First, I _____ the situation using only the most important words. Then, I choose a _____ to represent the _____ quantity. Last, I translate the phrase into an _____ .
2. What is a two-step expression?	an algebraic _____ containing two _____ operations

Summary

How can writing phrases as algebraic expressions help me solve problems?

_____
_____
_____
_____
_____
_____

Problem-Solving Investigation
Act It Out

Case 3 Teams

Twenty-four students will be divided into **four equal-size teams**.

Each student will count off, beginning with the number 1 as the first team.

If Nate is the **eleventh student** to count off, to which team number will he be assigned?

- Understand:

- Plan:

- Solve:

- Check:

Case 4 Savings

Dakota has **$5.38 in her savings account**.

Each week she **adds $2.93**.

How much money does Dakota have after **5 weeks**? after *n weeks*?

- Understand:

- Plan:

- Solve:

- Check:

Lesson 5 Vocabulary

Algebra: Properties

Use the three column chart to organize the vocabulary in this lesson. Write the word in Spanish. Then write the definition of each word.

English	Spanish	Definition
properties		
Commutative Property		
Associative Property		
equivalent expressions		
Identity Properties		

Inquiry Lab Guided Writing

The Distributive Property

HOW can you use models to evaluate and compare expressions?

Use the exercises below to help answer the Inquiry Question. Write the correct word or phrase on the lines provided.

1. Rewrite the question in your own words.

_____

_____

2. What key words do you see in the question?

_____

3. A _____ helps you see relationships between values.

4. To _____ an expression means "to find the value of an expression."

5. The expression $3(2 + 4)$ is represented below by what kind of a model? _____

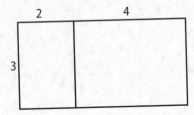

6. What word describes expressions that have the same value? _____

7. Use algebra tiles to represent the expressions $3(3x + 1)$ and $9x + 3$.

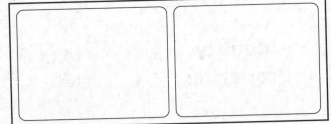

8. What does the model show you about the expressions?

_____

HOW can you use models to evaluate and compare expressions?

_____

_____

_____

Lesson 6 Notetaking

The Distributive Property

Use Cornell notes to better understand the lesson's concepts. Complete each answer by filling in the blanks with the correct word or phrase.

Questions	Notes
1. How do I use the Distributive Property?	I use the Distributive Property to _____ a sum by a number. I _____ each addend by the number outside the _____ .
2. How do I factor an expression?	I write each term of the expression using _____ and identify the common factors. I rewrite each term using the _____ . Then I use the _____ to write the expression as a product of the factors.

Summary
How can the Distributive Property help me to rewrite expressions?
_____
_____
_____
_____
_____
_____
_____

Inquiry Lab Guided Writing

Equivalent Expressions

HOW do you know that two expressions are equivalent?

Use the exercises below to help answer the Inquiry Question.
Write the correct word or phrase on the lines provided.

1. Rewrite the question in your own words.

_____

_____

2. What key words do you see in the question?

_____

3. Two expressions that have the same value are _____ .

Use the expressions $2x + 5x + 8$ and $7x + 6 + 2$ to answer Exercises 4-8.

4. How many *x*-tiles are needed to model the first expression? _____

5. How many *x*-tiles are needed to model the second expression? _____

6. How many 1-tiles are needed to model the first expression? _____

7. How many 1-tiles are needed to model the second expression? _____

8. Are the expressions equivalent? _____

HOW do you know that two expressions are equivalent?

_____

_____

Lesson 7 Vocabulary
Equivalent Expressions

Use the word cards to define each vocabulary word or phrase.

Word Cards

term	término

Definition

_____

_____

_____

Definición

_____

_____

_____

Circle the terms in the expression below.

$$5x + 3y - 6$$

- -

Word Cards

coefficient	coeficiente

Definition

_____

_____

_____

Definición

_____

_____

_____

Circle the coefficients in the terms below.

$$2z \quad 7p \quad -10y$$

Lesson 1 Vocabulary

Equations

Use the vocabulary squares to write a definition, a sentence, and an example for each vocabulary word.

equation	Definition
Example	**Sentence**

equals sign	Definition
Example	**Sentence**

solve	Definition
Example	**Sentence**

Inquiry Lab Guided Writing

Solve and Write Addition Equations

HOW do you solve addition equations using models?

Use the exercises below to help answer the Inquiry Question.
Write the correct word or phrase on the lines provided.

1. Rewrite the question in your own words.

_____

_____

2. What key words do you see in the question?

_____

3. What operation is used to combine, or add, numbers? _____

Use the equation 12 + *h* = 25 and the bar diagram below for Exercises 4-8.

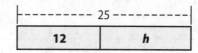

4. What is the unknown? _____

5. What is the other addend? _____

6. What does the full length of the bar diagram represent? _____

7. What related operation could you use to solve the equation? _____

8. Write a subtraction sentence shown by the bar diagram: _____

HOW do you solve addition equations using models?

_____

_____

_____

Lesson 2 Vocabulary
Solve and Write Addition Equations

Use the word cards to define each vocabulary word or phrase and give an example.

Word Cards

inverse operations

Definition

_____

_____

_____

Example Sentence

_____

_____

operaciones inversas

Definición

_____

_____

_____

Word Cards

Subtraction Property of Equality

Definition

_____

_____

_____

Example Sentence

_____

_____

propiedad de sustracción de la igualdad

Definición

_____

_____

_____

Inquiry Lab Guided Writing

Solve and Write Subtraction Equations

HOW do you solve subtraction equations using models?

Use the exercises below to help answer the Inquiry Question.
Write the correct word or phrase on the lines provided.

1. Rewrite the question in your own words.

_____

_____

2. What key words do you see in the question?

_____

3. What operation is used for taking away part of a whole? _____

4. What models could you use to show subtraction? _____

Use the bar diagram below for Exercises 5-7.

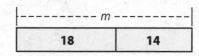

5. What is the total amount shown on the bar diagram? _____

6. What are the two parts shown on the bar diagram? _____

7. Write two subtraction sentences that are represented by the bar diagram.

_____

_____

HOW do you solve subtraction equations using models?

_____

_____

_____

Lesson 3 Notetaking

Solve and Write Subtraction Equations

Use Cornell notes to better understand the lesson's concepts. Complete each sentence by filling in the blanks with the correct word or phrase.

Questions	Notes
1. How can I solve a subtraction equation?	I can use _____ to solve a subtraction equation, because subtraction and _____ are _____ .
2. What does the Addition Property of Equality say I can do to an equation?	I can _____ the _____ number to each side of an equation and the sides will remain _____ .

Summary

How can the Addition Property of Equality be used to solve subtraction equations?

_____

_____

_____

_____

_____

_____

Problem-Solving Investigation
Guess, Check, and Revise

Case 3 Quizzes

On a science quiz, Ivan earned **18 points**.

There are **six problems worth 2 points** each and **two problems worth 4 points** each.

Find the number of problems of each type Ivan answered correctly.

- Understand:

- Plan:

- Solve:

- Check:

Case 4 Numbers

Kathryn is thinking of **four numbers** from **1 through 9** with a **sum of 18**.

Each number is used only once. Find the numbers.

- Understand:

- Plan:

- Solve:

- Check:

Inquiry Lab Guided Writing

Solve and Write Multiplication Equations

HOW do you solve multiplication equations using models?

Use the exercises below to help answer the Inquiry Question.
Write the correct word or phrase on the lines provided.

1. Rewrite the question in your own words.

_____

_____

2. What key words do you see in the question?

_____

3. In a multiplication equation, the total amount is the _____ . The parts that are

multiplied are the _____ .

Use the bar diagram below for Exercises 4-7.

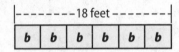

4. What is the total amount shown on the bar diagram? _____

5. What is the unknown factor in the bar diagram? _____

6. How can you use the bar diagram to find the other factor in the equation?

_____

_____

7. Write two multiplication sentences that are represented by the bar diagram.

_____   _____

HOW do you solve multiplication equations using models?

_____

_____

_____

Lesson 4 Vocabulary

Solve and Write Multiplication Equations

Use the definition map to list qualities about the vocabulary word or phrase.

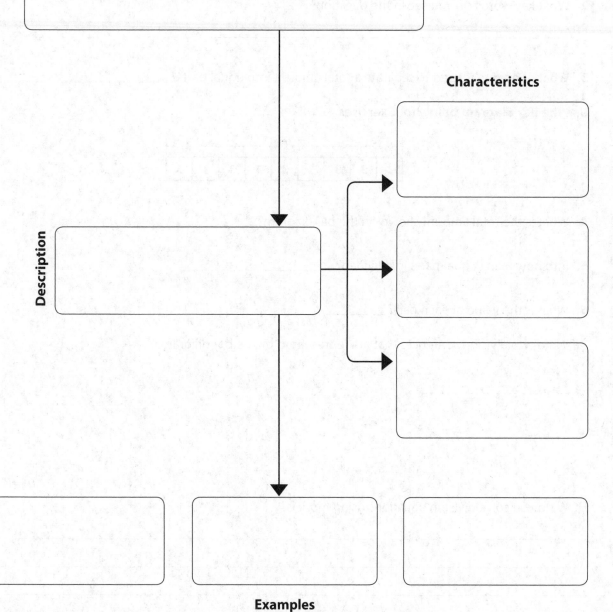

Vocabulary

Division Property of Equality

Description

Characteristics

Examples

Inquiry Lab Guided Writing

Solve and Write Division Equations

HOW do you solve division equations using models?

Use the exercises below to help answer the Inquiry question.
Write the correct word or phrase on the lines provided.

1. Rewrite the question in your own words.

_____

_____

2. What key words do you see in the question?

_____

3. What operation is used to separate a total amount into equal parts? _____

Use the bar diagram below for Exercises 4-7.

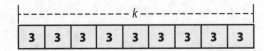

4. What is the total amount shown on the bar diagram? _____

5. Into how many equal parts is *k* divided? _____

6. What is the value of each part? _____

7. Write two division sentences that are represented by the bar diagram.

_____

_____

HOW do you solve division equations using models?

_____

_____

_____

Lesson 5 Notetaking

Solve and Write Division Equations

Use Cornell notes to better understand the lesson's concepts. Complete each
answer by filling in the blanks with the correct word or phrase.

Questions	Notes
1. How can I solve a division equation?	I can use _____ to solve for a division equation, because division and _____ are _____ .
2. What does the Multiplication Property of Equality say I can do to an equation?	I can _____ each side of an equation by the _____ nonzero number, and the sides will remain _____ .

Summary

When solving an equation, why is it necessary to perform the same operation on
each side of the equals sign?

_____

_____

_____

_____

_____

_____

Lesson 1 Vocabulary

Function Tables

Use the three column chart to organize the vocabulary in this lesson.
Write the word in Spanish. Then write the definition of each word.

English	Spanish	Definition
function		
function rule		
function table		
independent variable		
dependent variable		

Lesson 2 Vocabulary
Function Rules

Use the vocabulary squares to write a definition, a sentence, and an example for each vocabulary word.

	Definition
sequence	
Example	**Sentence**

	Definition
arithmetic sequence	
Example	**Sentence**

	Definition
geometric sequence	
Example	**Sentence**

Lesson 3 Vocabulary
Functions and Equations

**Use the concept web to identify different characteristics of a linear function.
Use a graph in one of the pieces of the web.**

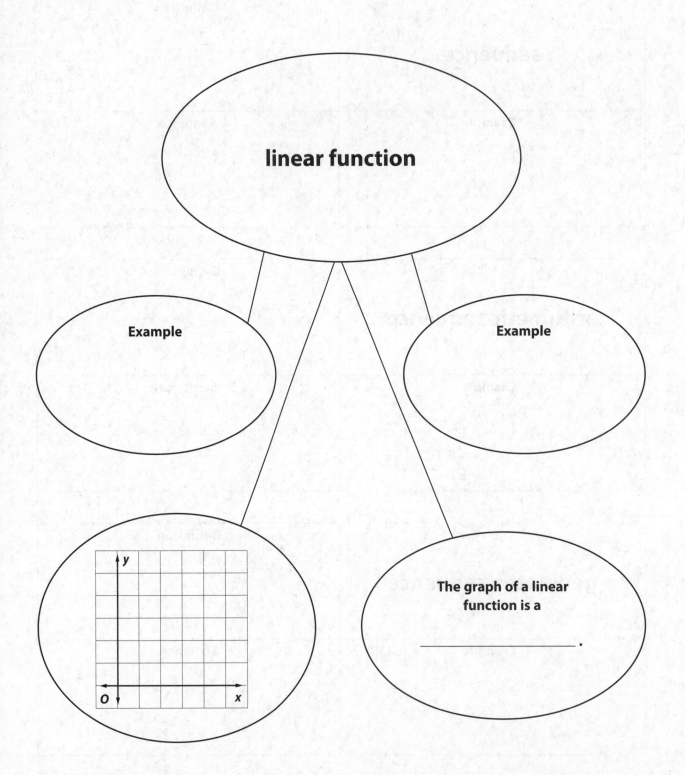

linear function

Example

Example

The graph of a linear
function is a

_____ .

Lesson 4 Notetaking

Multiple Representations of Functions

Use the definition map to list qualities about the multiple representations of the function.

Words

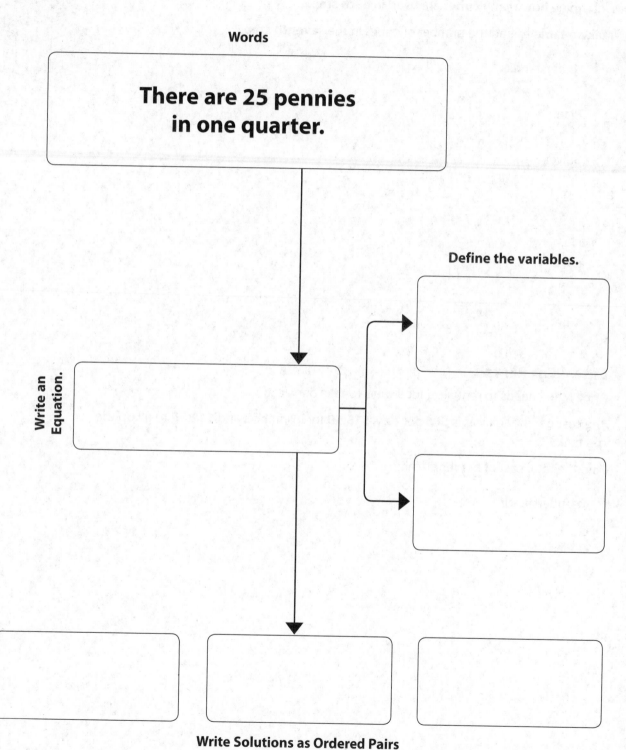

There are 25 pennies
in one quarter.

Define the variables.

Write an Equation.

Write Solutions as Ordered Pairs

Problem-Solving Investigation
Make a Table

Case 3 Geometry

Determine **how many cubes** are **used** in **each step.**

Make a table to find the number of cubes in the **seventh step**.

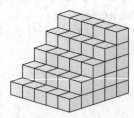

- Understand:

- Plan:

- Solve:

- Check:

Case 4 Car Rental

Anne Marie needs to rent a car for **9 days** to take on vacation.

The cost of renting a car is **$66 per day, $15.99** for insurance, and **$42.50** to fill up the gas tank.

Find the total cost of her rental car.

- Understand:

- Plan:

- Solve:

- Check:

Inquiry Lab Guided Writing

Inequalities

HOW can bar diagrams help you to compare quantities?

**Use the exercises below to help answer the Inquiry Question.
Write the correct word or phrase on the lines provided.**

1. Rewrite the question in your own words.

_____

_____

2. What key words do you see in the question?

_____

3. What does the > symbol mean? _____

4. What does the < symbol mean? _____

Use the bar diagrams below to answer the Exercises 5-7.

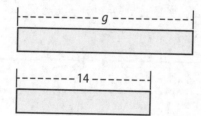

5. Is *g* greater than or less than 14? _____

6. How can you tell?

_____

7. Write the inequality shown by the bar diagram: _____

HOW can bar diagrams help you to compare quantities?

_____

_____

Lesson 5 Vocabulary

Inequalities

Use the word cards to define each vocabulary word or phrase and give an example.

Word Cards

inequality

Definition

_____

_____

_____

Example Sentence

_____

_____

desigualdad

Definición

_____

_____

_____

Word Cards

variable

Definition

_____

_____

_____

Example Sentence

_____

_____

variable

Definición

_____

_____

_____

Lesson 6 Review Vocabulary

Write and Graph Inequalities

Use the concept web to show examples of inequalities using words and symbols.

Words

Symbols

Words

Symbols

inequality

Symbols

Words

Symbols

Words

Inquiry Lab Guided Writing

Solve One-Step Inequalities

HOW can you use bar diagrams to solve one-step inequalities?

Use the exercises below to help answer the Inquiry question.
Write the correct word or phrase on the lines provided.

1. Rewrite the question in your own words.

_____

_____

2. What key words do you see in the question?

_____

3. A math sentence that compares quantities is called an _____ .

4. What symbols are used to show an inequality? _____

Use the bar diagram below to answer Exercises 5-8:

5. What is the given value? _____

6. What value of x would make the total amount equal to 10? _____

7. What value of x would make the total amount greater than 10? Write the inequality.

_____

8. What value of x would make the total amount less than 10? Write the inequality.

_____

HOW can you use bar diagrams to solve one-step inequalities?

_____

_____

_____

Lesson 7 Notetaking

Solve One-Step Inequalities

Use Cornell notes to better understand the lesson's concepts. Complete each sentence by filling in the blanks with the correct word or phrase.

Questions	Notes
1. How do I use Addition and Subtraction Properties to solve inequalities?	I can _____ or _____ the same number from each _____ of an inequality and the inequality remains _____.
2. How do I use Multiplication and Division Properties to solve inequalities?	I can _____ or _____ the same _____ number from each _____ of an inequality and the inequality remains _____ .

Summary

How is solving an inequality similar to solving an equation?

_____

_____

_____

_____

_____

_____

_____

Inquiry Lab Guided Writing

Area of Parallelograms

HOW does finding the area of a parallelogram relate to finding the area of a rectangle?

Use the exercises below to help answer the Inquiry Question.
Write the correct word or phrase on the lines provided.

1. Rewrite the question in your own words.

_____

_____

2. What key words do you see in the question?

_____

For Exercises 3 and 4, write the name of each figure on the line provided.

3.

_____

4.

_____

5. The number of square units needed to cover the inside of a figure is its _____ .

6. What operation is used to find the areas of rectangles and parallelograms?

_____

7. To find the area of a rectangle, multiply _____ and _____ .

8. To find the area of a parallelogram, multiply _____ and

_____ .

HOW does finding the area of a parallelogram relate to finding the area of a rectangle?

_____

_____

_____

Lesson 1 Vocabulary
Area of Parallelograms

Use the three column chart to organize the vocabulary in this lesson. Write the word in Spanish. Then write the definition of each word.

English	Spanish	Definition
polygon		
parallelogram		
rhombus		
base base		
height height		
formula $A = bh$ $A = \ell w$		

Inquiry Lab Guided Writing

Area of Triangles

HOW can you use the area of a parallelogram to find the area of a triangle?

Use the exercises below to help answer the Inquiry Question.
Write the correct word or phrase on the lines provided.

1. Rewrite the question in your own words.

_____

_____

2. What key words do you see in the question?

_____

3. You multiply the base and height to find the area of a _____ .

4. The dotted line divides this parallelogram into two equal _____ .

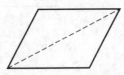

5. The area of the parallelogram above is 40 square units. What is the area of one triangle?

_____

6. How do you know?

_____

_____

7. Write a number sentence that shows how to find the area of the triangle.

_____

_____

HOW can you use the area of a parallelogram to find the area of a triangle?

_____

_____

Lesson 2 Vocabulary
Area of Triangles

Use the definition map to list qualities about the vocabulary word or phrase.

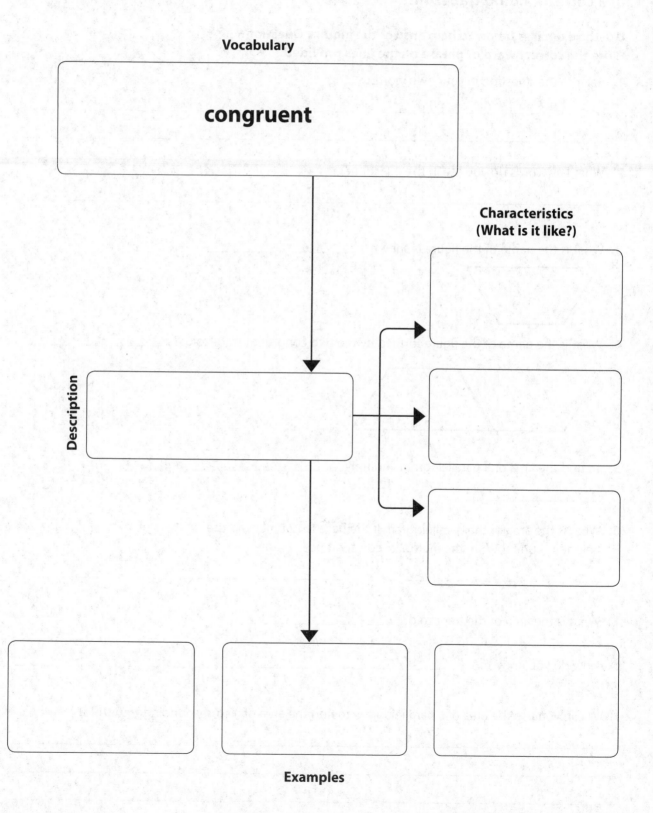

Vocabulary

congruent

**Characteristics
(What is it like?)**

Description

Examples

Inquiry Lab Guided Writing

Area of Trapezoids

HOW can you use the area of a parallelogram to find the area of a corresponding trapezoid?

Use the exercises below to help answer the Inquiry Question.
Write the correct word or phrase on the lines provided.

1. Rewrite the question in your own words.

   _____

   _____

2. What key words do you see in the question?

   _____

   _____

3. What is the name of the figure is shown? _____

4. What is the name of the figure formed by the two congruent trapezoids? _____

5. To find the area of a parallelogram, multiply _____ and

   _____ .

6. What is the area of the parallelogram if the base is 12 units and the
   height is 4 units? Write the multiplication sentence.

   _____

7. What is the area of the trapezoid? _____

8. How do you know? _____

   HOW can you use the area of a parallelogram to find the area of a corresponding trapezoid?

   _____

   _____

Lesson 3 Review Vocabulary

Area of Trapezoids

Use the concept web to name the characteristics of the shape.

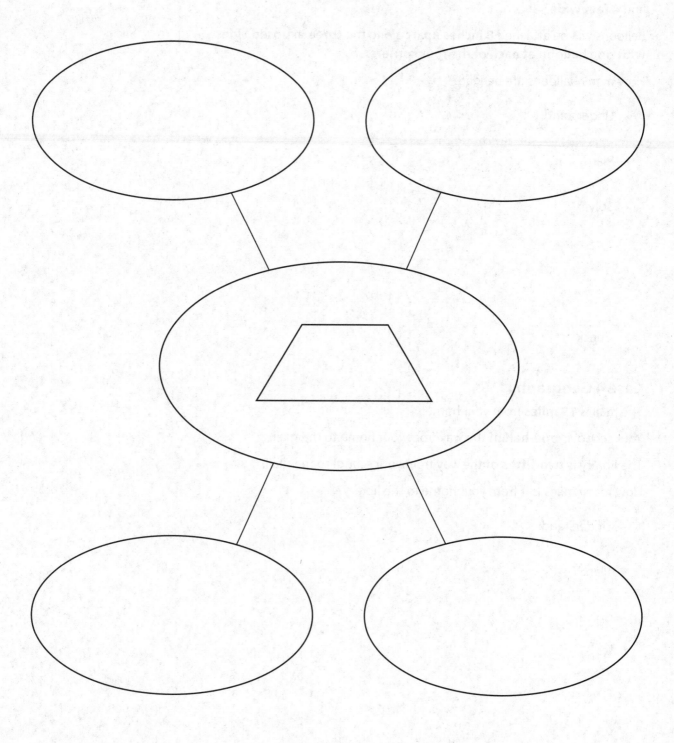

Problem-Solving Investigation
Draw a Diagram

Case 3 Decorations

A rectangular table that is placed lengthwise against a wall is **8 feet long** and **4 feet wide**.

Balloons will be attached **8 inches apart** along the **three exposed sides**, **with one balloon at each of the four corners.**

How many balloons are needed?

- Understand:

- Plan:

- Solve:

- Check:

Case 4 Geography

The mall is **15 miles** from your home.

Your school is **one-half of** the way from your home to the mall.

The library is **two-fifths of** the way from your school to the mall.

How many miles is it from your home to the library?

- Understand:

- Plan:

- Solve:

- Check:

Lesson 4 Review Vocabulary

Changes in Dimension

Use the definition map to list qualities about the vocabulary word or phrase.

Vocabulary

perimeter

**Characteristics
(What is it like?)**

Description

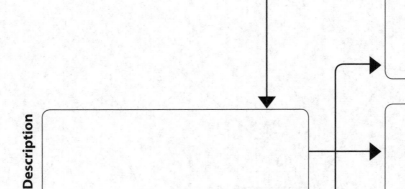

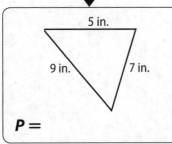

P =

P =

P =

Find the perimeter of each figure.

Lesson 5 Review Vocabulary

Polygons on the Coordinate Plane

Use the concept web to identify the parts of the coordinate plane.

Word Bank		
ordered pair	*x*-axis	*y*-axis
origin	*x*-coordinate	*y*-coordinate

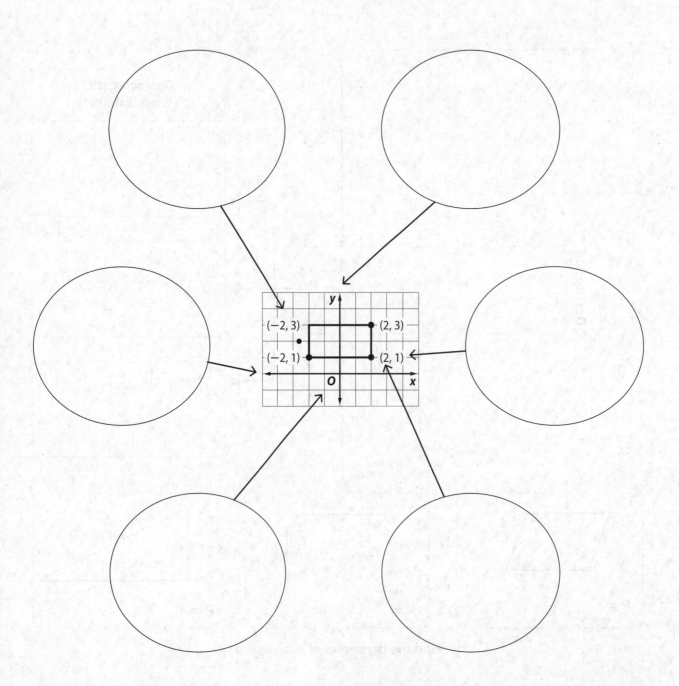

Inquiry Lab Guided Writing
Area of Irregular Figures

HOW can you estimate the area of an irregular figure?

**Use the exercises below to help answer the Inquiry Question.
Write the correct word or phrase on the lines provided.**

1. Rewrite the question in your own words.

_____

_____

2. What key words do you see in the question?

_____

3. A number that tells about how much is an _____ .

4. A figure that does not have a regular shape is an _____ .

5. An irregular figure is shown below. What two regular shapes can you see

   in the figure? _____

6. How can you estimate the area of the mushroom?

_____

7. Why can't you find the exact area of the mushroom?

_____

_____

HOW can you estimate the area of an irregular figure?

_____

_____

_____

Lesson 6 Notetaking

Area of Composite Figures

Use Cornell notes to better understand the lesson's concepts. Complete each sentence by filling in the blanks with the correct word or phrase.

Questions	Notes
1. How do I find the area of a composite figure?	I can _____ the composite shape into figures with _____ I know how to find. Then I can add those _____ .
2. How do I find the area of an overlapping figure?	First I find the _____ of each overlapping shape. Then I _____ those areas. Next I find the _____ of the overlapping section. Then I _____ the overlapping area.

Summary

How can you decompose figures to find area?

_____

_____

_____

_____

_____

_____

_____

Inquiry Lab Guided Writing

Volume of Rectangular Prisms

HOW can you use models to find volume?

Use the exercises below to help answer the Inquiry Question.
Write the correct word or phrase on the lines provided.

1. Rewrite the question in your own words.

_____

_____

2. What key words do you see in the question?

_____

3. _____ is the amount of space inside a three-dimensional figure.

4. What three dimensions are multiplied to find the volume of a rectangular prism?

_____

5. Volume is measured in _____ units.

6. What type of three-dimensional figure is best for modeling area? _____

7. What are some items shaped like a cube?

_____

8. If 16 centimeter cubes are used to make a rectangular prism, what is

the area of the figure? _____

HOW can you use models to find volume?

_____

_____

_____

Lesson 1 Vocabulary
Volume of Rectangular Prisms

Use the three column chart to organize the vocabulary in this lesson.
Write the word in Spanish. Then write the definition of each word.

English	Spanish	Definition
three-dimensional figure		
prism		
rectangular prism		
volume		
cubic units		

Lesson 2 Vocabulary
Volume of Triangular Prisms

Use the concept web to show what you know about the three-dimensional shape shown. Name the shape and write the formula used to find the volume in two of the pieces of the web.

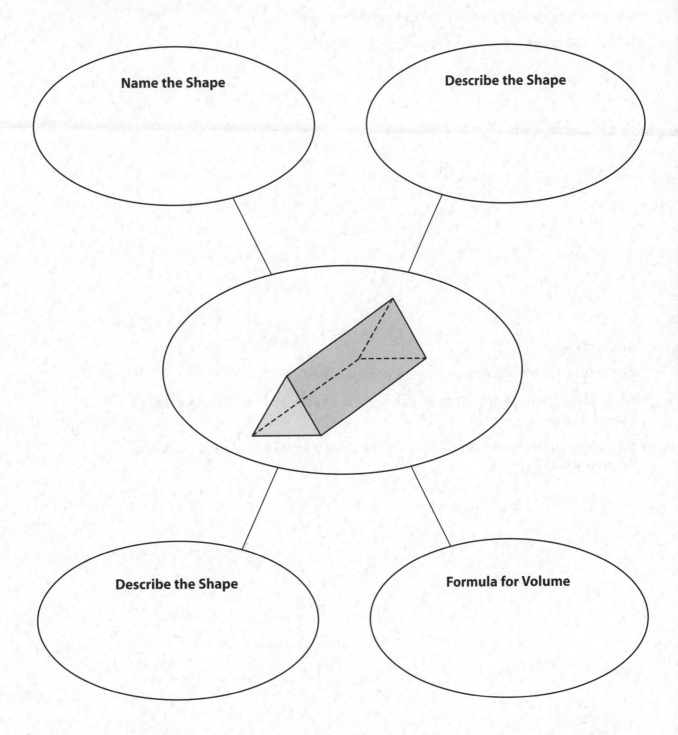

Name the Shape

Describe the Shape

Describe the Shape

Formula for Volume

Problem-Solving Investigation
Make a Model

Case 3 Assembly

DJ is helping set up **7 rows** of chairs for a school assembly.

There are **eight chairs in the first row**.

Each row after that has **two more chairs** than the previous row.

If he **has 100 chairs**, can he set up enough rows? Explain.

- Understand:

- Plan:

- Solve:

- Check:

Case 4 Paper

Timothy took a piece of notebook paper and **cut it in half**.

Then he placed the 2 pieces on top of each other and **cut them in half again** to have 4 pieces of paper.

If he could keep cutting the paper in this manner, how many pieces of paper would he have **after 6 cuts**?

- Understand:

- Plan:

- Solve:

- Check:

Inquiry Lab Guided Writing

Surface Area of Rectangular Prisms

HOW can you use nets to find surface area?

Use the exercises below to help answer the Inquiry Question.
Write the correct word or phrase on the lines provided.

1. Rewrite the question in your own words.

_____

_____

2. What key words do you see in the question?

_____

3. A _____ is a two-dimensional pattern of a three-dimensional figure.

4. Each flat side of a three-dimensional figure is called a _____ .

5. Name some real-life examples of rectangular prisms.

_____

_____

6. How many faces does a rectangular prism have? _____

7. You used a net to find the area of each face of a rectangular prism.
How would you find the total surface area of the rectangular prism?

_____

_____

HOW can you use nets to find surface area?

_____

_____

_____

Lesson 3 Vocabulary

Surface Area of Rectangular Prisms

Use the definition map to list qualities about the vocabulary word or phrase.

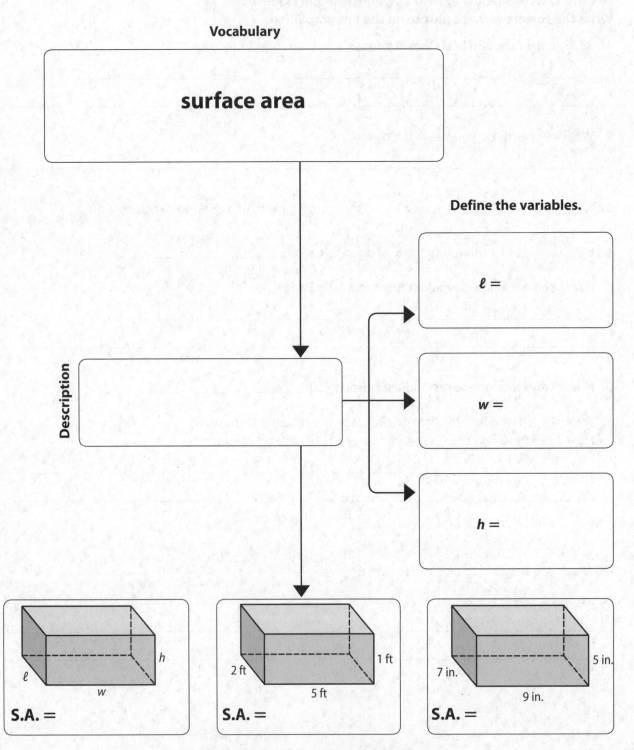

Vocabulary

surface area

Description

Define the variables.

$\ell =$

$w =$

$h =$

S.A. =

S.A. =

S.A. =

Find the surface area of each rectangular prism.

Inquiry Lab Guided Writing

Nets of Triangular Prisms

HOW is the area of a triangle related to the surface area of a triangular prism?

Use the exercises below to help answer the Inquiry Question.
Write the correct word or phrase on the lines provided.

1. Rewrite the question in your own words.

   _____

   _____

2. What key words do you see in the question?

   _____

3. What is a net?

   _____

   _____

4. How many faces does a triangular prism have? _____

5. What shape are the bases of a triangular prism? _____

6. How do you find the area of a triangle? _____

7. What shape are the other faces of a triangular prism? _____

8. How do you find the area of a rectangle? _____

   HOW is the area of a triangle related to the surface area of a triangular prism?

   _____

   _____

   _____

Lesson 4 Review Vocabulary

Surface Area of Triangular Prisms

Use the concept web to find the surface area of the triangular prism. Identify the shape of each face. Find the area of each face. Then find the total surface area.

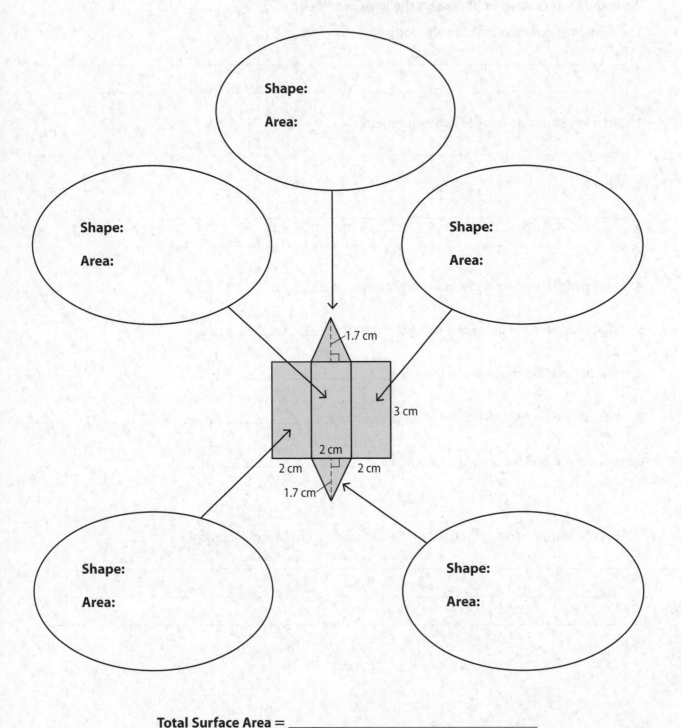

Shape:

Area:

Shape:

Area:

Shape:

Area:

Shape:

Area:

Shape:

Area:

1.7 cm

3 cm

2 cm

2 cm   2 cm

1.7 cm

Total Surface Area = _____

Inquiry Lab Guided Writing

Nets of Pyramids

HOW is the area of a triangle related to the surface area of a square pyramid?

Use the exercises below to help answer the Inquiry Question.
Write the correct word or phrase on the lines provided.

1. Rewrite the question in your own words.

_____

_____

2. What key words do you see in the question?

_____

_____

3. How many faces does a square pyramid have? _____

4. What shape is the base of a square pyramid? _____

5. How do you find the area of a square? _____

6. What shape are the other faces of a square pyramid? _____

7. How do you find the area of a triangle? _____

HOW is the area of a triangle related to the surface area of a square pyramid?

_____

_____

_____

Lesson 5 Vocabulary
Surface Area of Pyramids

Use the definition map to list qualities about the vocabulary word or phrase.

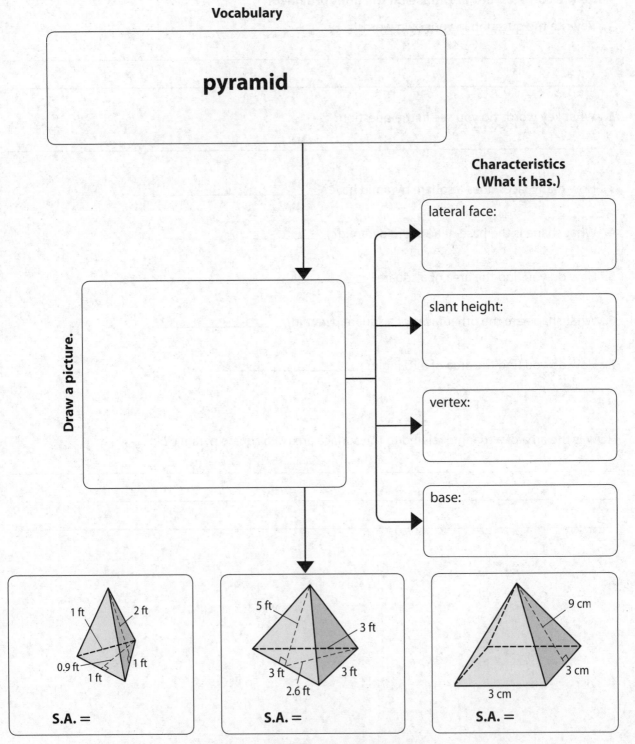

Vocabulary

pyramid

Draw a picture.

Characteristics (What it has.)

lateral face:

slant height:

vertex:

base:

S.A. =

S.A. =

S.A. =

Find the surface area of each pyramid.

Inquiry Lab Guided Writing

Statistical Questions

HOW are surveys created to collect and analyze data?

Use the exercises below to help answer the Inquiry Question. Write the correct word or phrase on the lines provided.

1. Rewrite the question in your own words.

   _____

   _____

2. What key words do you see in the question?

   _____

3. A _____ is a question or set of questions used to collect information.

4. Pieces of information are called _____ .

5. What kind of questions are used in surveys?

   _____

   _____

6. Write an example of a statistical question about food.

   _____

   _____

7. Data with a wide range of values can be separated into _____ .

8. How do you find the equal share of a set of data?

   _____

   _____

HOW are surveys created to collect and analyze data?

_____

_____

_____

_____

Lesson 1 Vocabulary
Mean

Use the definition map to list qualities about the vocabulary word or phrase.

Vocabulary

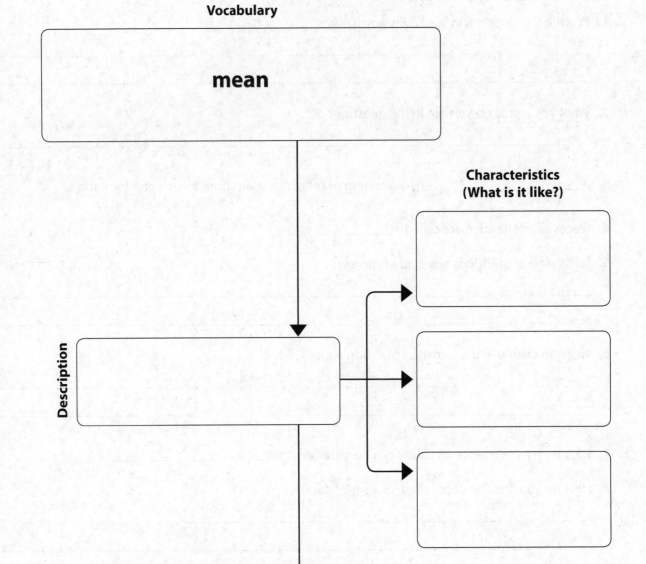

mean

Description

**Characteristics
(What is it like?)**

Examples

{7, 3, 6, 4, 5}

mean =

{27, 29, 30, 30}

mean =

{10, 12, 17, 13}

mean =

Lesson 2 Vocabulary
Median and Mode

Use the vocabulary squares to write a definition, a sentence, and an example for each vocabulary word.

	Definition
measures of center	
Find the measures of center for the following data set. {3, 4, 5, 5, 7, 9, 11, 12}	Sentence

	Definition
median	
Find the median of the following data set. {3, 4, 5, 5, 7, 9, 11, 12}	Sentence

	Definition
mode	
Find the mode of the following data set. {3, 4, 5, 5, 7, 9, 11, 12}	Sentence

Problem-Solving Investigation
Use Logical Reasoning

Case 3 Marketing

A survey showed that **70** customers bought **white bread, 63** bought **wheat bread,** and **35** bought **rye bread.**

Of those who bought exactly two types of bread, **12** bought **wheat and white, 5** bought white and rye, and **7** bought **wheat and rye.**

Two customers **bought all three.**

How many customers bought only wheat bread?

- Understand:

- Plan:

- Solve:

- Check:

Case 4 Pets

Dr. Poston is a veterinarian.

One week she treated **20 dogs, 16 cats**, and **11 birds**.

Some owners had more than one pet, as shown in the table.

How many owners had **only a dog** as a pet?

Pet	Number of Owners
dog and cat	7
dog and bird	5
cat and bird	3
dog, cat and bird	2

- Understand:

- Plan:

- Solve:

- Check:

Lesson 3 Vocabulary

Measures of Variation

Use the three column chart to organize the vocabulary in this lesson. Write the word in Spanish. Then write the definition of each word.

English	Spanish	Definition
measures of variation		
quartiles		
first quartile		
third quartile		
interquartile range		
range		
outliers		

Lesson 4 Notetaking

Mean Absolute Deviation

**Use Cornell notes to better understand the lesson's concepts. Complete each
sentence by filling in the blanks with the correct word or phrase.**

Questions	Notes
1. How do I find the mean absolute deviations for a data set?	I find the _____ of the distances between each data value and the _____ , then I _____ by the number of data values.
2. How do I compare mean absolute deviations for two data sets?	The data set with a _____ mean absolute deviation has data values that are _____ to the mean than a data set with a _____ mean absolute deviation.

Summary

What does the mean absolute deviation tell you about a set of data?

_____

_____

_____

_____

_____

_____

Lesson 5 Review Vocabulary
Appropriate Measures

Use the concept web to describe the data set using the measures of center.

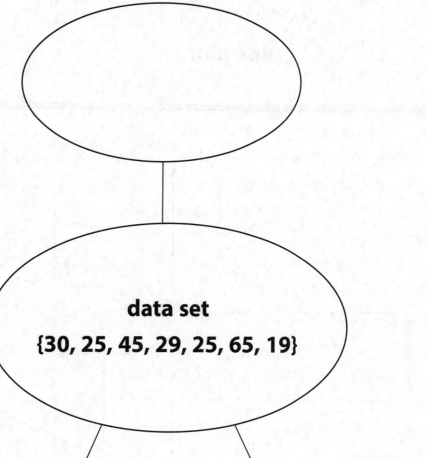

data set

{30, 25, 45, 29, 25, 65, 19}

Which measure of center best describes the data set? _____

Lesson 1 Vocabulary
Line Plots

Use the definition map to list qualities about the vocabulary word or phrase.

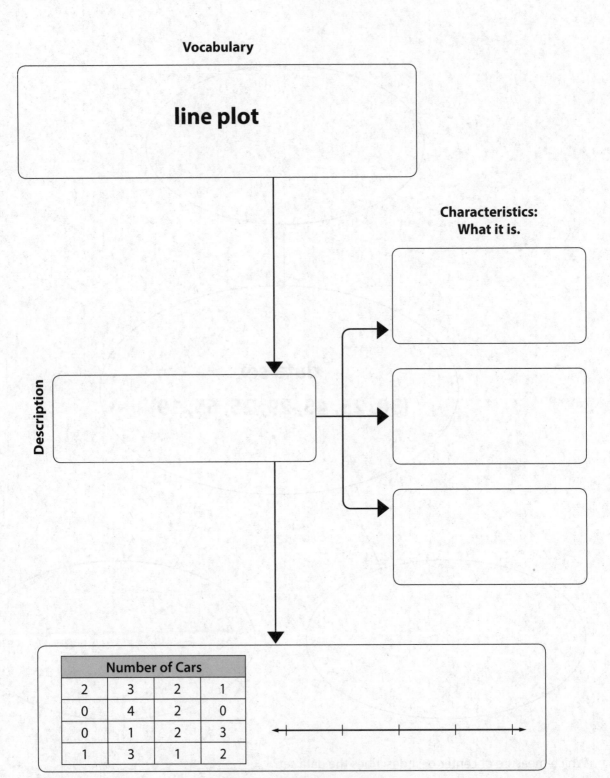

Vocabulary

line plot

Description

**Characteristics:
What it is.**

Number of Cars			
2	3	2	1
0	4	2	0
0	1	2	3
1	3	1	2

Draw a line plot for the given data.

Lesson 2 Vocabulary
Histograms

Use the word cards to define each vocabulary word or phrase and give an example.

Word Cards

histogram	**histograma**
Definition	**Definición**
_____	_____
_____	_____
_____	_____
Example Sentence	
_____	_____
_____	_____

Word Cards

frequency distribution	**distribución de frecuencias**
Definition	**Definición**
_____	_____
_____	_____
_____	_____
Example Sentence	
_____	_____
_____	_____

Lesson 3 Review Vocabulary

Box Plots

Use the concept web to identify the parts of a box plot.

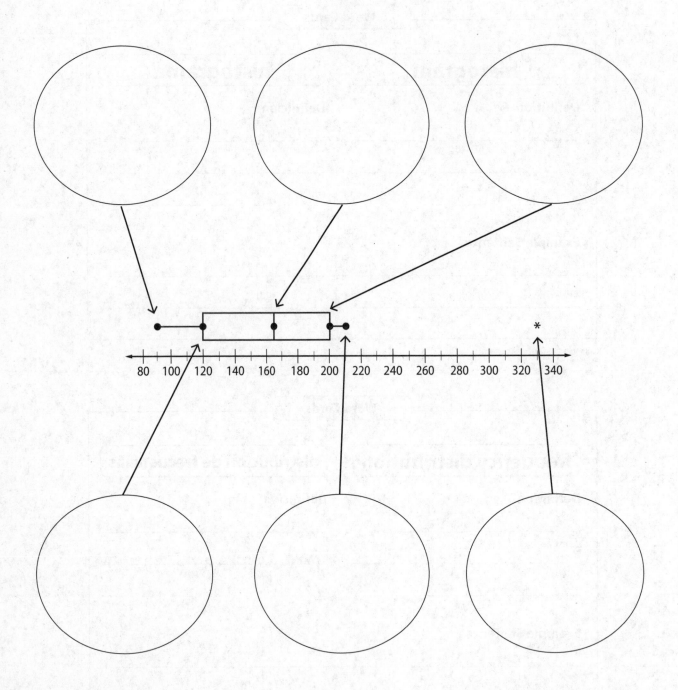

Problem-Solving Investigation
Use a Graph

Case 3 Lawn Mowing

DeShawn mowed lawns over the summer to earn extra money.

The number of lawns he mowed each week is shown in the line plot.

What is the **mean number of lawns he mowed**?

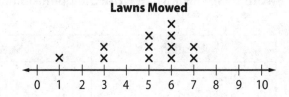

- Understand:

- Plan:

- Solve:

- Check:

Case 4 Magazines

The **box plot** shows the **number of magazines sold** for a club fundraiser.

What is the **difference** between the **median number** of magazines sold and **the most** magazines sold?

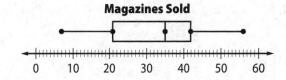

- Understand:

- Plan:

- Solve:

- Check:

Lesson 4 Vocabulary
Shape of Data Distributions

Use the three column chart to organize the vocabulary in this lesson. Write the word in Spanish. Then write the definition of each word.

English	Spanish	Definition
distribution		
symmetric distribution		
cluster		
gap		
peak		

Inquiry Lab Guided Writing

Collect Data

HOW do you answer a statistical question?

Use the exercises below to help answer the Inquiry question. Write the correct word or phrase on the lines provided.

1. Rewrite the question in your own words.

_____

_____

2. What key words do you see in the question?

_____

3. When do you use a statistical question?

_____

4. Fill in the blanks to explain the steps of answering a statistical question:

  a. Create a plan to collect _____ .

  b. Conduct a _____ to collect data from a group of people.

  c. Create a table or graph to _____ the data.

5. Name some ways you can display data.

_____

_____

HOW do you answer a statistical question?

_____

_____

_____

Lesson 5 Vocabulary
Interpret Line Graphs

Use the definition map to list qualities about the vocabulary word or phrase.

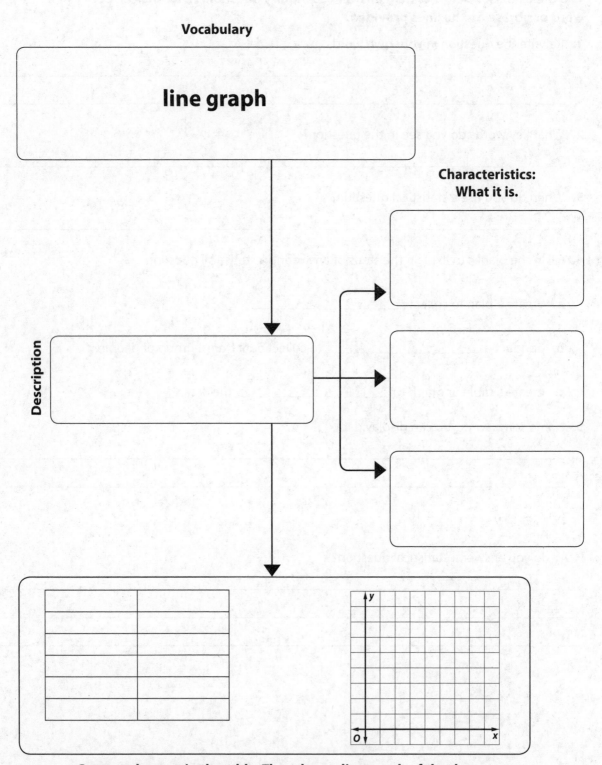

Vocabulary

line graph

Characteristics:
What it is.

Description

Create a data set in the table. Then draw a line graph of the data.

Lesson 6 Review Vocabulary

Select an Appropriate Display

Use the three column chart to organize the vocabulary in this lesson. Write the word in Spanish. Then write the definition of each word.

English	Spanish	Definition
bar graph		
box plot		
histogram		
line graph		
line plot		

Inquiry Lab Guided Writing

Use Appropriate Units and Tools

HOW do you determine a measureable attribute?

Use the exercises below to help answer the Inquiry Question.
Write the correct word or phrase on the lines provided.

1. Rewrite the question in your own words.

_____

_____

2. What key words do you see in the question?

_____

_____

3. Name some features of a box that you can measure.

_____

4. What tool would you use to measure length? _____

5. What tool would you use to measure weight? _____

For Exercises 6-8, list units you could use to measure each attribute.

6. capacity or volume _____

7. length _____

8. weight or mass _____

9. Would you use inches, feet, or yards to record the length of a pencil? _____

10. What interval would you use for a graph scale that shows lengths of pencils?

_____

HOW do you determine a measureable attribute?

_____

_____

_____

What are VKVs® and How Do I Create Them?

Visual Kinesthetic Vocabulary Cards® are flashcards that animate words by focusing on their structure, use, and meaning. The VKVs in this book are used to show cognates, or words that are similar in Spanish and English.

Step 1

Go to the back of your book to find the VKVs for the chapter vocabulary you are currently studying. Follow the cutting and folding instructions at the top of the page. The vocabulary word on the BLUE background is written in English. The Spanish word is on the ORANGE background.

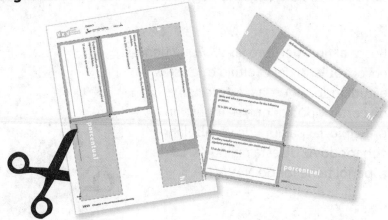

Step 2

There are exercises for you to complete on the VKVs. When you understand the concept, you can complete each exercise. All exercises are written in English and Spanish. You only need to give the answer once.

Step 3

Individualize your VKV by writing notes, sketching diagrams, recording examples, and forming plurals (radius: radii or radiuses).

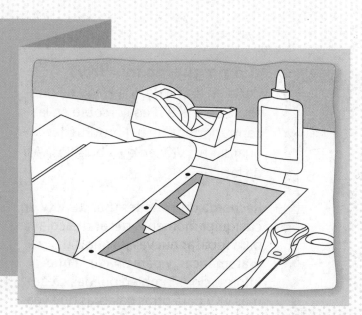

How Do I Store My VKVs?

Take a 6" x 9" envelope and cut away a V on one side only. Glue the envelope into the back cover of your book. Your VKVs can be stored in this pocket!

Remember you can use your VKVs ANY time in the school year to review new words in math, and add new information you learn. Why not create your own VKVs for other words you see and share them with others!

Las tarjetas de vocabulario visual y cinético (VKV) contienen palabras con animación que está basada en la estructura, uso y significado de las palabras. Las tarjetas de este libro sirven para mostrar cognados, que son palabras similares en español y en inglés.

Paso 1

Busca al final del libro las VKV que tienen el vocabulario del capítulo que estás estudiando. Sigue las instrucciones de cortar y doblar que se muestran al principio. La palabra de vocabulario con fondo AZUL está en inglés. La de español tiene fondo NARANJA.

Paso 2

Hay ejercicios para que completes con las VKV. Cuando entiendas el concepto, puedes completar cada ejercicio. Todos los ejercicios están escritos en inglés y español. Solo tienes que dar la respuesta una vez.

Paso 3

Da tu toque personal a las VKV escribiendo notas, haciendo diagramas, grabando ejemplos y formando plurales (radio: radios).

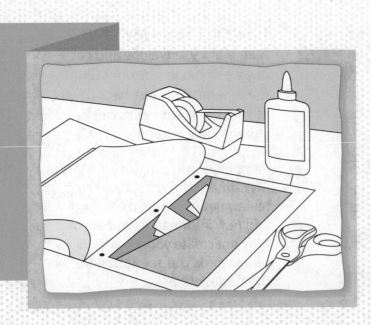

¿Cómo guardo mis VKV?

Corta en forma de "V" el lado de un sobre de 6" X 9". Pega el sobre en la contraportada de tu libro. Puedes guardar tus VKV en esos bolsillos. ¡Así de fácil!

Recuerda que puedes usar tus VKV en cualquier momento del año escolar para repasar nuevas palabras de matemáticas, y para añadir la nueva información. También puedes crear más VKV para otras palabras que veas, y poder compartirlas con los demás.

Dinah Zike's
V K V Visual
Kinesthetic
Vocabulary ®

Chapter 1

✂ cut on all dashed lines   🗀 fold on all solid lines

Define graph. (Define graficar.)

graph

coordinate plane

Dinah Zike's
Visual
Kinesthetic
Vocabulary®

VKV

Chapter 1

✂ cut on all dashed lines

▭ fold on all solid lines

aficar

plano de coordenadas

Draw a coordinate plane on the other side. Then label the parts of the coordinate plane. (Dibuja un plano de coordenadas en el otro lado. Luego, rotula sus partes.)

Graph the ordered pairs. (Haz una gráfica de los pares ordenados.) (3, 8), (4, 10), (5, 5)

Dinah Zike's
VKV
Visual
Kinesthetic
Vocabulary ®

Chapter 1

✂ cut on all dashed lines   fold on all solid lines

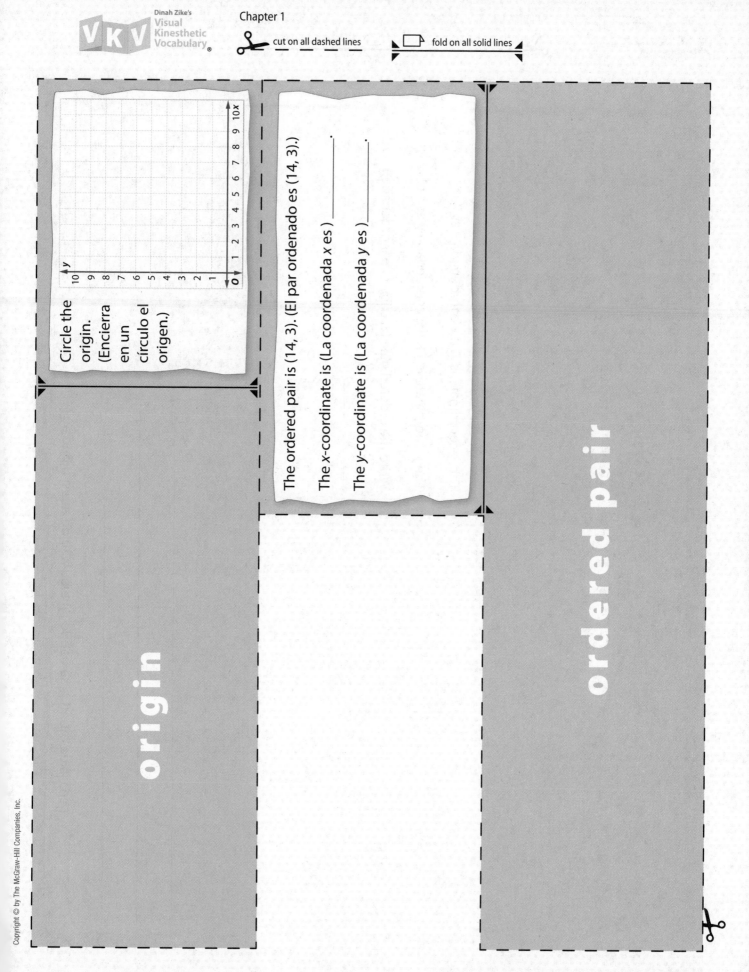

Circle the origin. (Encierra en un círculo el origen.)

The ordered pair is (14, 3). (El par ordenado es (14, 3).)

The x-coordinate is (La coordenada x es) _____ .

The y-coordinate is (La coordenada y es) _____ .

origin

ordered pair

par ordenado

en

The x-coordinate is 7. (La coordenada x es 7.)

The y-coordinate is 11. (La coordenada y es 11.)

Write the ordered pair. (Escribe el par ordenado.)

(,)

The origin is the point where (El origen es el punto en el cual)

Dinah Zike's
V K V Visual
Kinesthetic
Vocabulary®

Chapter 1

✂ cut on all dashed lines   fold on all solid lines

Circle the greater unit price. (Encierra en un círculo el mayor precio unitario.)

6 tickets for $84

5 tickets for $75

unit price

y -coordinate

x

Define unit price. (Define precio unitario.)

_____

_____

_____

Circle the y-coordinates (Encierra en un círculo las coordenadas y.)

(4, 18)   (23, 12)

(5.9, 65)   (0.9, 12.1)

Dinah Zike's
Visual
Kinesthetic
Vocabulary®

Chapter 1

✂ cut on all dashed lines

📁 fold on all solid lines

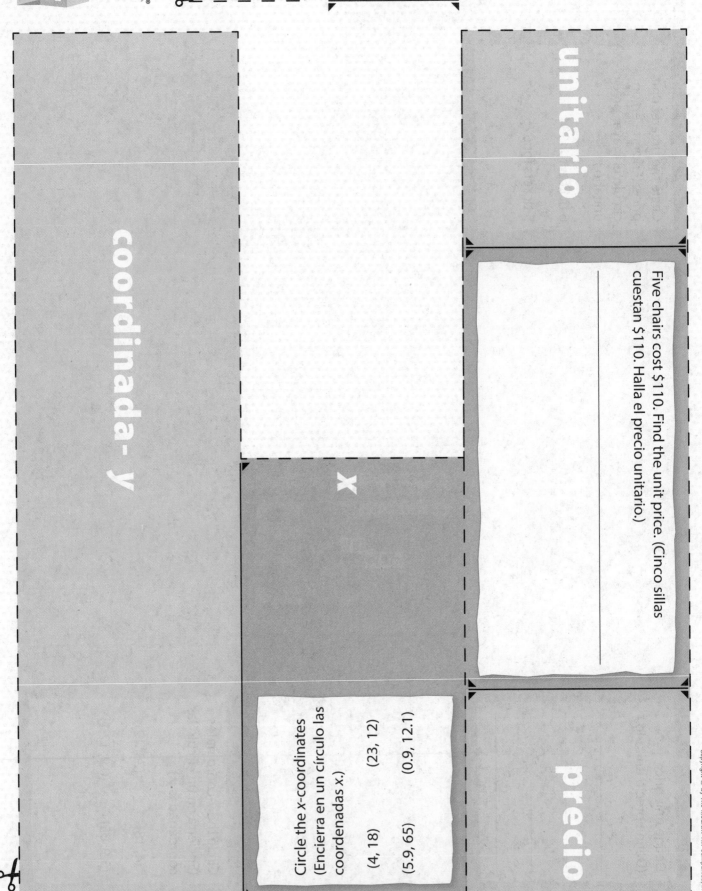

unitario

coordinada-y

x

Five chairs cost $110. Find the unit price. (Cinco sillas cuestan $110. Halla el precio unitario.)

_____

precio

Circle the x-coordinates
(Encierra en un círculo las
coordenadas x.)

(4, 18)   (23, 12)

(5.9, 65)   (0.9, 12.1)

Dinah Zike's
Visual
Kinesthetic
Vocabulary®

✂ cut on all dashed lines   ⬚ fold on all solid lines

List 3 different forms of rational numbers. (Enumera tres formas de representar los números racionales.)

Write about a time when it might be useful to know the percent proportion. (Escribe acerca de una situación en la cual podría ser útil conocer la proporción portcentual.)

rational number

percent proportion

Write each number as a fraction (Escribe los números como fracciones.)

12 = _____

0.35 = _____

3.1 = _____

Write the percent proportion. (Escribe la proporción porcentual.)

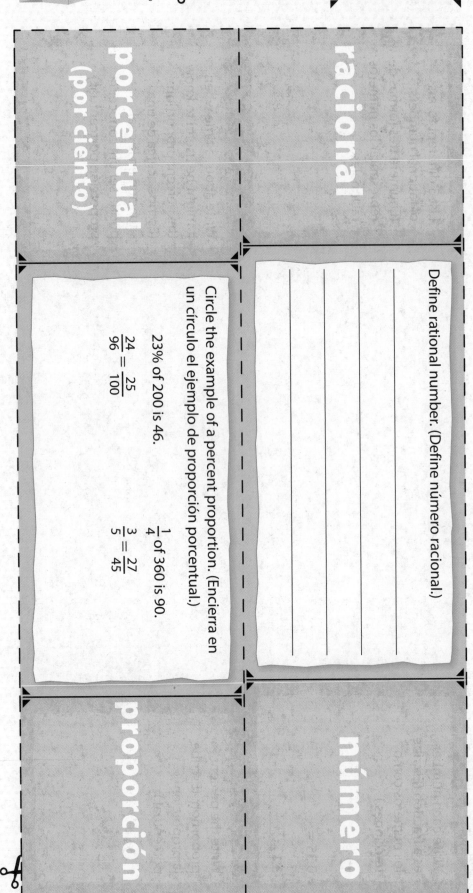

porcentual
(por ciento)

racional

Define rational number. (Define número racional.)

Circle the example of a percent proportion. (Encierra en un círculo el ejemplo de proporción porcentual.)

23% of 200 is 46.

$\frac{1}{4}$ of 360 is 90.

$\frac{24}{96} = \frac{25}{100}$

$\frac{3}{5} = \frac{27}{45}$

proporción

número

✂ cut on all dashed lines   fold on all solid lines

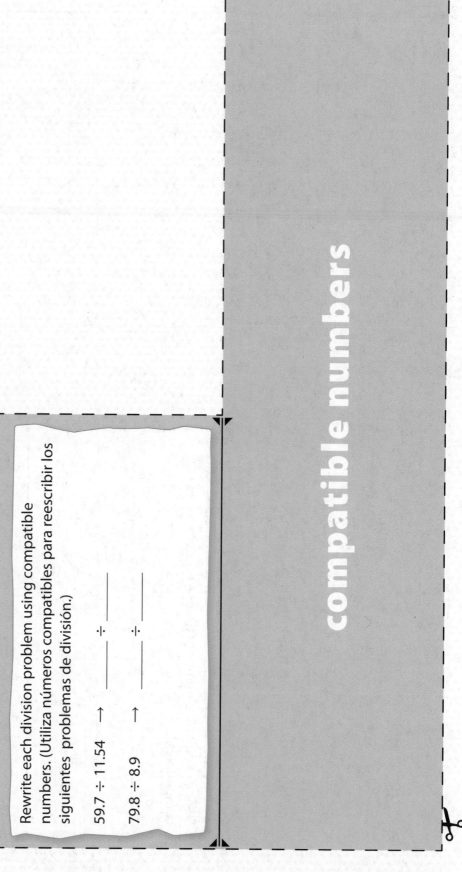

compatible numbers

Rewrite each division problem using compatible numbers. (Utiliza números compatibles para reescribir los siguientes problemas de división.)

59.7 ÷ 11.54   →   _____ ÷ _____

79.8 ÷ 8.9   →   _____ ÷ _____

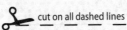
números compatibles

Define compatible numbers. (Define número compatibles.)

Dinah Zike's
VKV
Visual
Kinesthetic
Vocabulary®

Chapter 4

✂ cut on all dashed lines   ⬚ fold on all solid lines

Commutative Property

Circle the word that is related to the Commutative Property. (Encierra en un círculo la palabra que se relaciona con la propiedad conmutativa.)

grouping   order

Define Commutative Property. (Define propiedad conmutativa.)

reciprocal

To solve $8 \div \frac{3}{4}$, multiply _____ by the reciprocal of _____. (Para resolver $8 \div \frac{3}{4}$, multiplica _____ por el recíproco de _____.)

dimensional analysis

Define dimensional analysis. (Define análisis dimensional.)

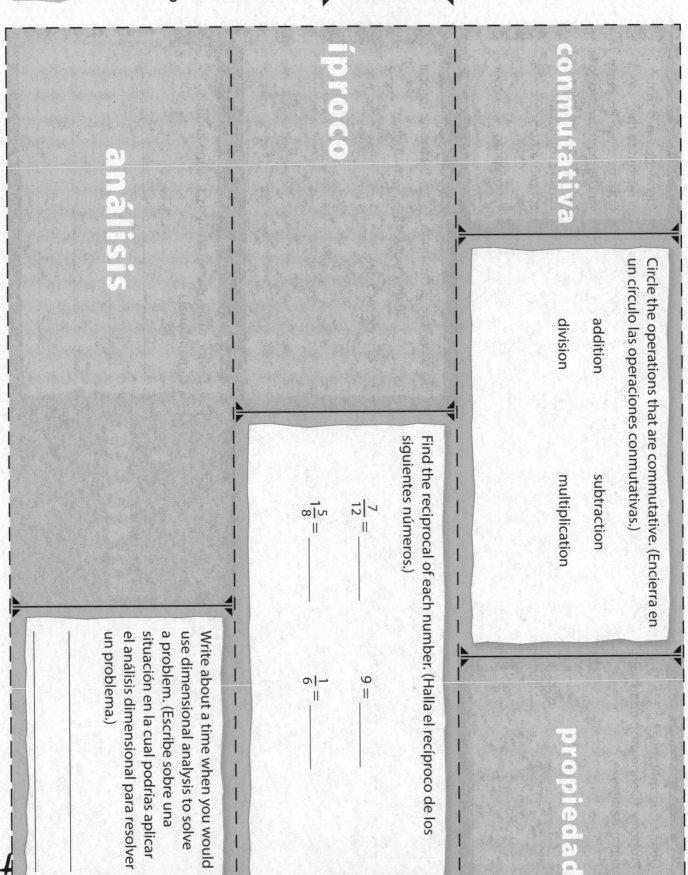

conmutativa

íproco

análisis

propiedad

Circle the operations that are commutative. (Encierra en un círculo las operaciones conmutativas.)

addition   subtraction

division   multiplication

Find the reciprocal of each number. (Halla el recíproco de los siguientes números.)

$\frac{7}{12} =$ _____

$1\frac{5}{8} =$ _____

$9 =$ _____

$\frac{1}{6} =$ _____

Write about a time when you would use dimensional analysis to solve a problem. (Escribe sobre una situación en la cual podrías aplicar el análisis dimensional para resolver un problema.)

_____

_____

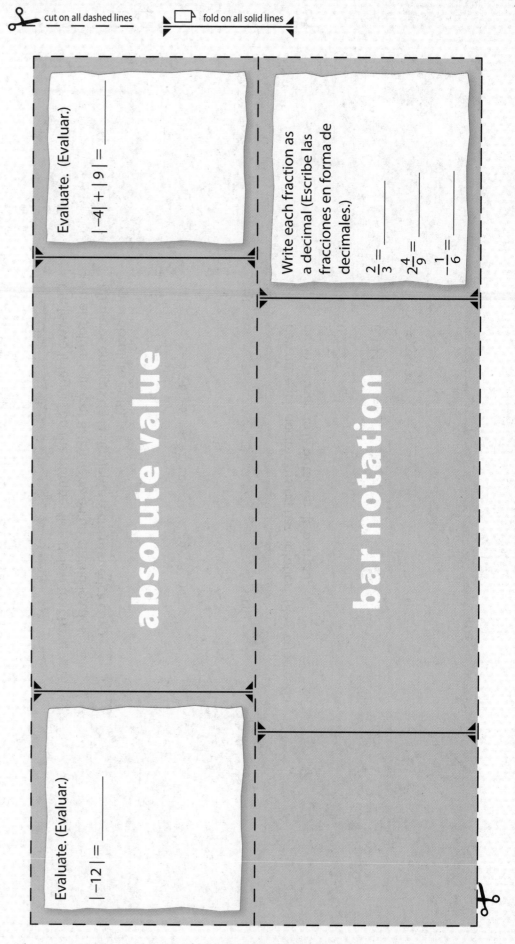

Evaluate. (Evaluar.)

$|-4| + |9| = $ _____

Write each fraction as a decimal (Escribe las fracciones en forma de decimales.)

$\dfrac{2}{3} = $ _____

$2\dfrac{4}{9} = $ _____

$-\dfrac{1}{6} = $ _____

absolute value

bar notation

Evaluate. (Evaluar.)

$|-12| = $ _____

Dinah Zike's
Visual
Kinesthetic
Vocabulary ®

✂ cut on all dashed lines

▱ fold on all solid lines

de barra

absoluto

Circle two integers on the number line below that have an absolute value of 8. (Encierra en un círculo los dos números enteros de la siguiente recta numérica cuyo valor absoluto es 8.)

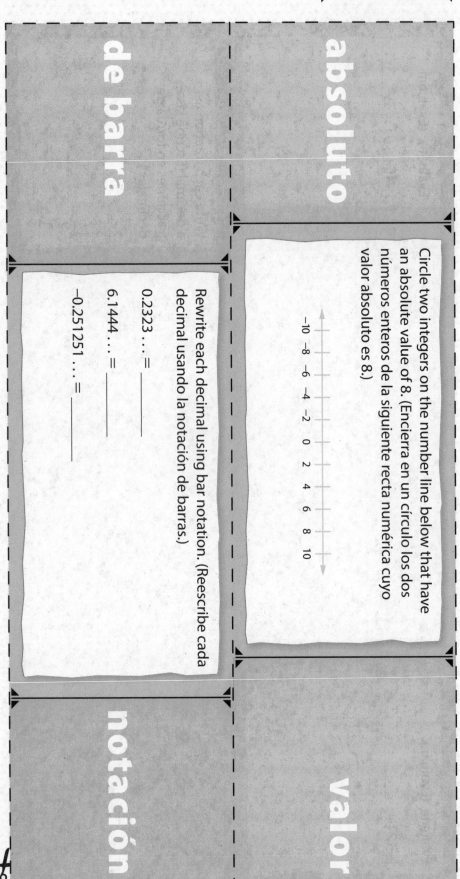

–10 –8 –6 –4 –2 0 2 4 6 8 10

Rewrite each decimal using bar notation. (Reescribe cada decimal usando la notación de barras.)

0.2323 . . . = ____

6.1444 . . . = ____

–0.251251 . . . = ____

notación

valor

✂ cut on all dashed lines

▭ fold on all solid lines

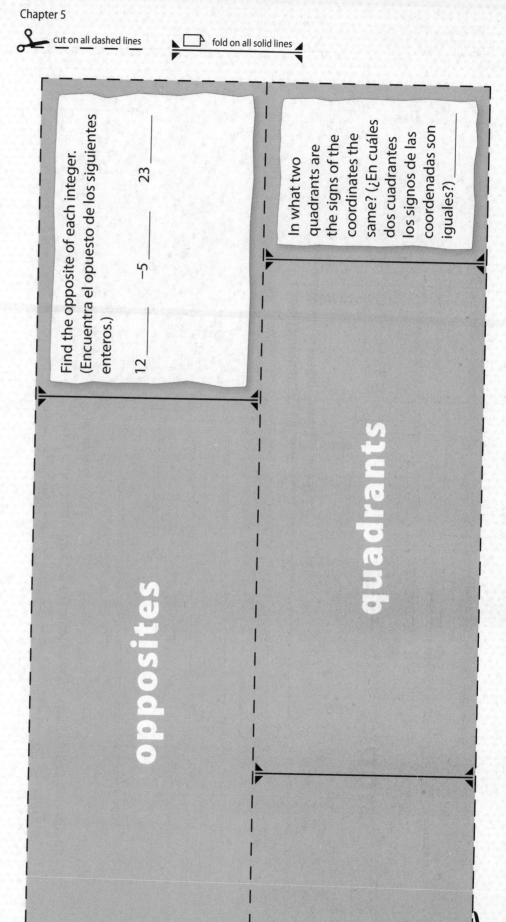

Find the opposite of each integer. (Encuentra el opuesto de los siguientes enteros.)

12 _____   −5 _____   23 _____

In what two quadrants are the signs of the coordinates the same? (¿En cuáles dos cuadrantes los signos de las coordenadas son iguales?) _____

opposites

quadrants

Dinah Zike's
VKV
Visual
Kinesthetic
Vocabulary®

Chapter 5

✂ cut on all dashed lines

☐ fold on all solid lines

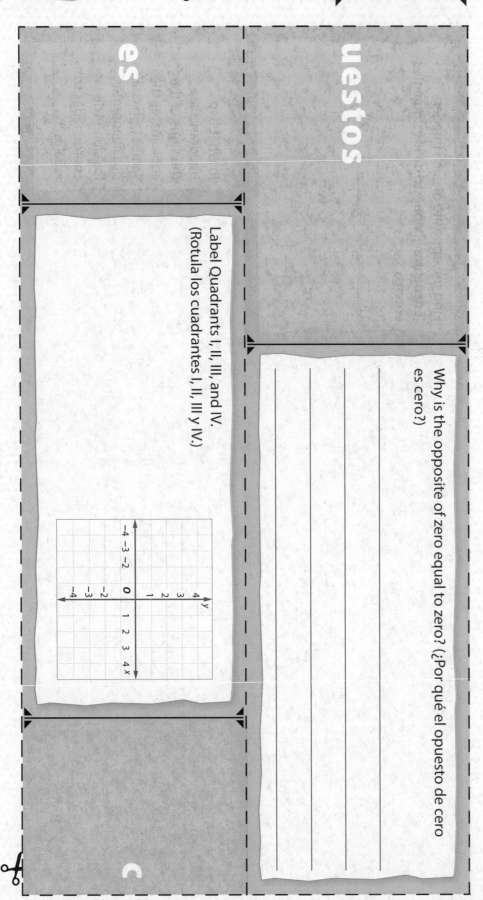

es

uestos

Label Quadrants I, II, III, and IV.
(Rotula los cuadrantes I, II, III y IV.)

Why is the opposite of zero equal to zero? (¿Por qué el opuesto de cero
es cero?)

C

cut on all dashed lines

fold on all solid lines

Define properties. (Define propiedades.)

Evaluate $12m$ if $m = 3$. (Evalúa $12m$ si $m = 3$.)

Define coefficient. (Define coeficiente.)

properties

evaluate

coefficient

Dinah Zike's
Visual
Kinesthetic
Vocabulary®

Chapter 6

✂ cut on all dashed lines

fold on all solid lines

iciente

r

iedades

Name the property that is shown in each example. (Nombra la propiedad que se muestra en los siguientes ejemplos.)

$56 + 0 = 56$

$12 \cdot 5 = 5 \cdot 12$

$3 + (6 + 4) = (3 + 6) + 4$

What information do you need to be able to evaluate $6x + 3$? (¿Qué información necesitas para evaluar la ecuación $6x + 3$?)

Circle the coefficients in the expressions below. (Encierra en un círculo los coeficientes de las siguientes expresiones.)

$6x + 3 = 21$

$15 - 2p$

$m - 3y$

$24 + 3a = 9$

Dinah Zike's
V K V
Visual
Kinesthetic
Vocabulary ®

Chapter 6

✂ cut on all dashed lines

◢▢◣ fold on all solid lines

Rewrite the expression $3x + x + x$ so that it has only one term. (Reescribe la expresión $3x + x + x$ de manera que tenga un único término.)

expression

term

numérica

Simplify each numerical expression. (Simplifica las siguientes expresiones numéricas.)

$3 + 12 - (2 \times 3) =$ _____

$15 \div 3 + 20 - 8 =$ _____

Dinah Zike's
VKV Visual
Kinesthetic
Vocabulary®

Chapter 6

✂ cut on all dashed lines

📄 fold on all solid lines

término

expresión

numerical

How many terms are in each expression? (¿Cuántos términos tienen las siguientes expresiones?)

x + 4 − 3y _____

9r + 4s − t _____

12 − 5a _____

m + m + m + m _____

Define numerical expression. (Define expresión numérica.)

Dinah Zike's
Visual
Kinesthetic
Vocabulary ®

Chapter 7

✂ cut on all dashed lines

fold on all solid lines

Define equation. (Define ecuación.)

Circle the solution to each equation. (Encierra en un círculo la solución de las siguientes ecuaciones.)

$27 - y = 9$ 14   16   18

$x + 52 = 100$ 47   48   49

$36 \div m = 12$ 2   3   4

What is the inverse operation of division? (¿Cuál es la operación inversa de la división?)

equation

solution

inverse operations

What is the inverse operation of addition? (¿Cuál es la operación inversa de la adición?)

Dinah Zike's
Visual
Kinesthetic
Vocabulary®

inversas

ción

cuación

operaciones

Solve each equation using inverse operations. (Utiliza las operaciones inversas para resolver las siguientes ecuaciones.)

$x + 12 = 25$

$x =$ _____

$16 = y - 3$

_____ $= y$

Define solution. (Define solución.)

Aiden bought 24 marbles for $6. Write and solve an equation to show how much each marble cost. (Andrés compró 24 canicas en $6. Escribe y resuelve una ecuación que muestre cuánto costó cada una.)

Dinah Zike's
Visual
Kinesthetic
Vocabulary ®

Chapter 8

✂ cut on all dashed lines

⬚ fold on all solid lines

In a function, the x-value is the _____ and the y-value is the _____. (En una función, x es la _____ e y es la _____.)

function table

Define function. (Define función.)

A linear function is a function that (Una función lineal es aquella que)

linear function

Dinah Zike's
Visual
Kinesthetic
Vocabulary®

Chapter 8

✂ cut on all dashed lines

📄 fold on all solid lines

funciones

función lineal

tabla de

Complete the function table. (Completa el tabla de función.)

Input (x)	3x	Output
0		
2		
5		

Write an equation to represent the function. (Escribe una ecuación que represente la función.)

Input (x)	1	2	3	4
Output (y)	0	4	8	12	16

_____

Dinah Zike's
Visual
Kinesthetic
Vocabulary®

Chapter 8

✂ cut on all dashed lines   ▢ fold on all solid lines

geometric sequence

aritmética

Circle the correct word. (Encierra en un círculo la palabra correcta.)

An arithmetic sequence uses (addition/multiplication). (Una sucesión geométrica usa la (adición/multiplicación).)

Dinah Zike's
VKV Visual
Kinesthetic
Vocabulary®

Chapter 8

✂ cut on all dashed lines

▭ fold on all solid lines

secesión geométrica

arithmetic

Circle the correct word. (Encierra en un círculo la palabra correcta.)

A geometric sequence uses (addition/multiplication). (Una progesión aritmética usa la (adición/multiplicación).)

✂ cut on all dashed lines   fold on all solid lines

Define congruent. (Define congruente.)

List three polygons. Draw one example. (Enumera tres polígonos. Dibuja un ejemplo.)

congruent

polygon

cut on all dashed lines   fold on all solid lines

Dinah Zike's
Visual
Kinesthetic
Vocabulary®

Chapter 9

✂ cut on all dashed lines

📋 fold on all solid lines

ígono

e

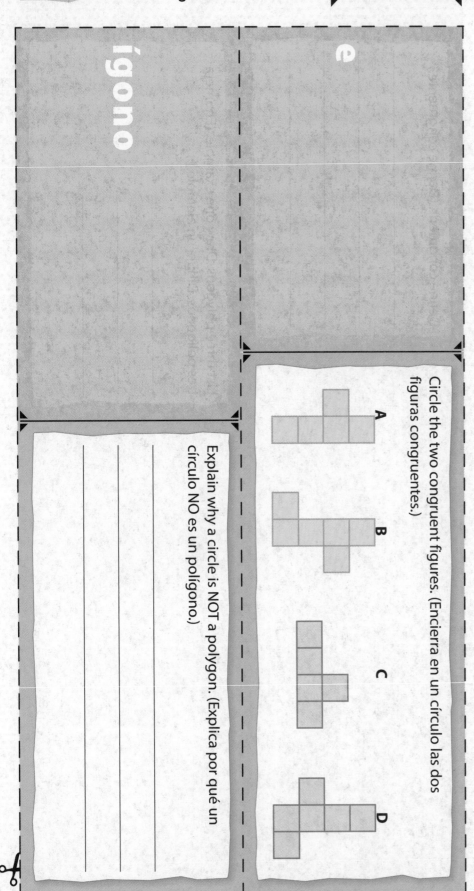

Circle the two congruent figures. (Encierra en un círculo las dos figuras congruentes.)

A

B

C

D

Explain why a circle is NOT a polygon. (Explica por qué un círculo NO es un polígono.)

Dinah Zike's
Visual
Kinesthetic
Vocabulary ®

Chapter 9

cut on all dashed lines

fold on all solid lines

Define rhombus. (Define rombo.)

Circle the formula that represents the area of a parallelogram. (Encierra en un círculo la fórmula que representa el área de un paralelogramo)

$A = b^2 + h^2$ $\qquad A = 2(b + h) \qquad A = bh$

rhombus

parallelogram

Dinah Zike's
VKV Visual Kinesthetic Vocabulary®

✂ cut on all dashed lines

◻ fold on all solid lines

paralelogramo

ombo

Draw a pair of parallel lines. (Dibuja un par de rectas paralelas.)

Explain why a square is a rhombus. (Explica por qué un cuadrado es un rombo.)

Dinah Zike's
Visual
Kinesthetic
Vocabulary®

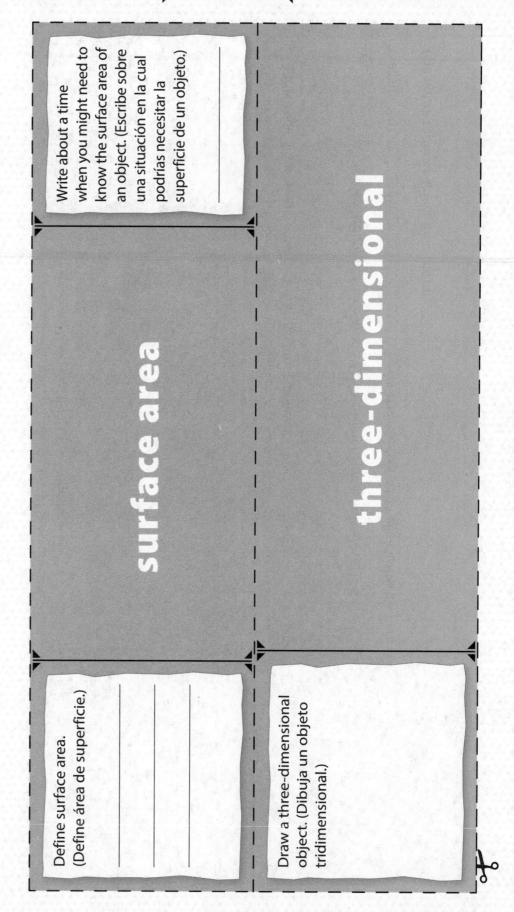

Write about a time when you might need to know the surface area of an object. (Escribe sobre una situación en la cual podrías necesitar la superficie de un objeto.)

surface area

three-dimensional

Define surface area. (Define área de superficie.)

Draw a three-dimensional object. (Dibuja un objeto tridimensional.)

Dinah Zike's
Visual
Kinesthetic
Vocabulary®

Chapter 10

✂ cut on all dashed lines

📄 fold on all solid lines

superficie

área de

tri.

Find the surface area. (Calcula la superficie.)

$S.A. = 2\ell h + 2\ell w + 2hw$

15 cm

7 cm

2 cm

$S.A. = $ _____

List the three dimensions. (Enumera las tres dimensiones.)

_____

_____

_____ is the amount of space inside a three-dimensional figure, and it is measured in _____. (_____ es el espacio que se halla dentro de una figura tridimensional. Se mide en _____ .)

Dinah Zike's
**Visual
Kinesthetic
Vocabulary**®

Chapter 10

✂ cut on all dashed lines

🗅 fold on all solid lines

Define *vertex*. (Define *vértice*.)

Draw a triangular prism.
(Dibuja un prisma
triangular.)

vertex

rectangular

triangular prism

Dinah Zike's
Visual
Kinesthetic
Vocabulary®

Draw a rectangular prism. (Dibuja un prisma rectangular.)

prisma triangular

 értice

rectangular

The plural of vertex is vertices. How many vertices does the figure have? (¿Cuántos vértices tiene la figura?)

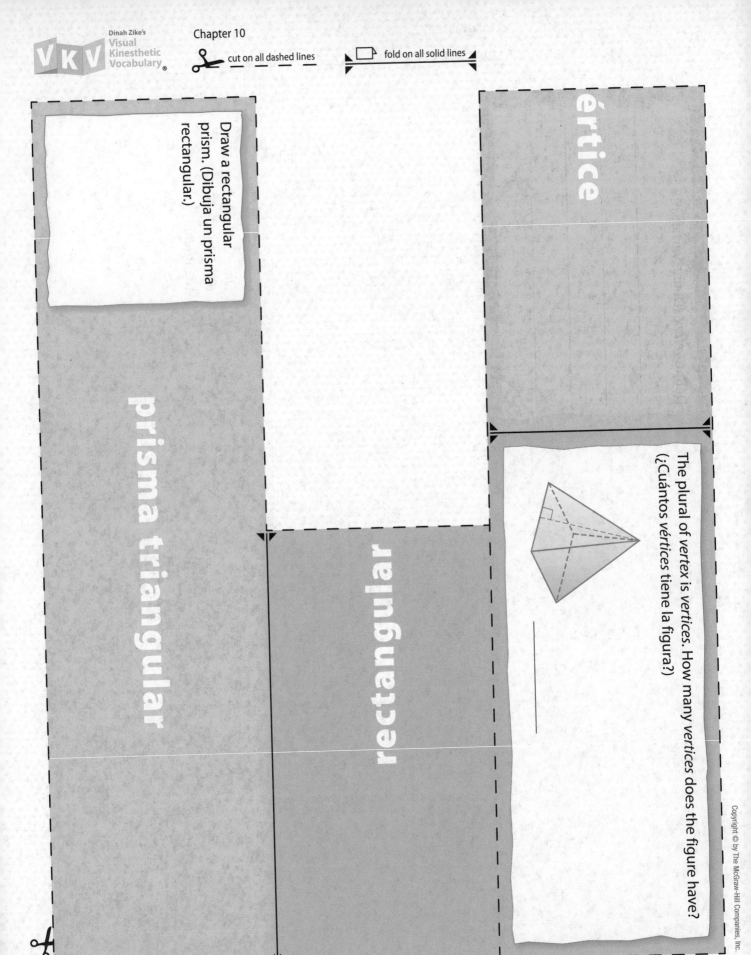

Dinah Zike's
Visual
Kinesthetic
Vocabulary®

Chapter 11

✂ cut on all dashed lines   fold on all solid lines

How is the median of a data set different from the mean? (¿Cuál es la diferencia entre la mediana y la media de un conjunto de datos?)

Define range. (Define rango.)

Define mode. (Define moda.)

median

range

mode

Dinah Zike's
Visual
Kinesthetic
Vocabulary®

cut on all dashed lines

fold on all solid lines

a

o

a

Find the median of the data set below. (Halla la mediana del siguiente conjunto de datos.)

| 12 | 15 | 10 | 18 | 12 | 14 | 13 | 17 | 18 |

Create a data set that has 6 numbers and a range of 12. (Crea una tabla de datos que tenga 6 números y un rango de 12.)

Find the mode of the data set below. (Halla la moda del siguiente conjunto de datos.)

| 2 | 2 | 3 | 1 | 4 | 2 | 3 | 1 | 2 |

Dinah Zike's
Visual Kinesthetic Vocabulary®

✂ cut on all dashed lines

✀ fold on all solid lines

interquartile

What is the interquartile range of the data set below?
(¿Cuál es el rango intercuartil del siguiente conjunto de datos?)

4   4   5   6   6   6   7   8   8   8

_____

Dinah Zike's
Visual
Kinesthetic
Vocabulary ®

Chapter 11

✂ cut on all dashed lines

fold on all solid lines

intercuartile

Circle the first and third quartiles in the data set below.
(Encierra en un círculo el primer y el tercer cuartiles en el siguiente conjunto de datos.)

4   4   5   6   6   6   7   8   8   8

Dinah Zike's
Visual
Kinesthetic
Vocabulary®

✂ cut on all dashed lines   📷 fold on all solid lines

frequency distribution

Write about a time when it might be useful to know the frequency distribution. (Escribe acerca de una situación en la cual podría ser útil conocer la distribución de frecuencias.)

Dinah Zike's
Visual
Kinesthetic
Vocabulary ®

Chapter 12

✂ cut on all dashed lines

▭ fold on all solid lines

distribución de frecuencias

Define distribution. (Define distribución.)

Dinah Zike's
Visual
Kinesthetic
Vocabulary®

Chapter 12

✂ --- --- --- cut on all dashed lines

▢ fold on all solid lines

If a data set has symmetric distribution, do you describe the center using the mean or median? Explain. (Si la distribución de un conjunto de datos es simétrica, ¿el centro se describe con la media o con la mediana? Explique.)

Define histogram. (Define histograma.)

symmetric

histogram

Dinah Zike's
Visual
Kinesthetic
Vocabulary®

Chapter 12

cut on all dashed lines

fold on all solid lines

a

imétrico

In the space at right, draw an example of a histogram. (En el espacio de la derecha dibuja un ejemplo de histograma.)

Which line plot shows symmetric distribution? (¿Cuál gráfica de puntos presenta una distribución simétrica?)

A

Number of States Visited

10 ×××
11 ×××××
12 ×××××××
13 ×××
14 ×
15
16
17
18
19 ×
20

B

Ages of Tennis Players (yr)

24 ×
25 ××
26 ×××
27 ×××
28 ××××
29 ××××
30 ××××
31 ×××
32 ××
33 ×
34